Aktuelle Forschung Medizintechnik – Latest Research in Medical Engineering

Editor-in-Chief:
Th. M. Buzug, Lübeck, Deutschland

Among future technologies with high innovation potential, medical engineering counts among those with above-average growth rates and is considered crises-proof. Computerization, miniaturization, and molecularization are essential trends in medical engineering. Computerization is the basis for medical imaging, image processing, and image-guided methods in surgery. Miniaturization plays an important role in the field of intelligent implants, minimally invasive surgery as well as in the development of new nanostructured materials in medicine. Molecularization is both a crucial element in the field of regenerative medicine and the so called molecular imaging. Cross-sectional technologies like nano- and microsystems technology as well as optical technologies and softwaresystems are, therefore, of high relevance.

This series for outstanding dissertations and habilitation treatises in the field of medical engineering covers clinical engineering and medical computer science as well as medical physics, biomedical engineering and medical engineering science.

Dagmar Kainmueller

Deformable Meshes for Medical Image Segmentation

Accurate Automatic Segmentation of Anatomical Structures

 Springer Vieweg

Dagmar Kainmueller
Max Planck Institute of Molecular Cell
  Biology and Genetics
Dresden, Germany

Dissertation University of Lübeck, 2013

ISBN 978-3-658-07014-4   ISBN 978-3-658-07015-1 (eBook)
DOI 10.1007/978-3-658-07015-1

The Deutsche Nationalbibliothek lists this publication in the Deutsche Nationalbibliografie;
detailed bibliographic data are available in the Internet at http://dnb.d-nb.de.

Library of Congress Control Number: 2014947962

Springer Vieweg
© Springer Fachmedien Wiesbaden 2015

Printed on acid-free paper

Springer Vieweg is a brand of Springer DE.
Springer DE is part of Springer Science+Business Media.
www.springer-vieweg.de

In memoriam Bernd Fischer

Preface by the Series Editor

The book Deformable Meshes for Accurate Automatic Segmentation of Medical Image Data by Dr. Dagmar Kainmüller is the 9th volume of the new Springer-Vieweg series of excellent theses in medical engineering. The thesis of Dr. Kainmüller has been selected by an editorial board of highly recognized scientists working in that field. The Springer-Vieweg series aims to establish a forum for Monographs and Proceedings on Medical Engineering. The series publishes works that give insights into the novel developments in that field. Prospective authors may contact the Series Editor about future publications within the series at:

Prof. Dr. Thorsten M. Buzug
Series Editor Medical Engineering

Institute of Medical Engineering
University of Lbeck
Ratzeburger Allee 160
23562 Lbeck
Web: www.imt.uni-luebeck.de
Email: buzug@imt.uni-luebeck.de

Foreword

"To Be — or not to Be: That is the question!"

This famous quote from 450 years old William Shakespeare from "The Tragicall Historie of Hamlet, Prince of Denmarke" (Act III, Scene I) is still vivid and vital and has not lost any of its power and magic. In fact, it can also be read as one of the most beautiful and charming descriptions of the art of segmentation: It manifests two antagonistic states and raises the fundamental question of how to distinguish these states. The ability to differentiate between good and bad or important and irrelevant is key in all matters of life.

The art of segmentation is central for medical imaging, where an exponential growth of imaging modalities and images can be observed. On the one hand, this information is very important and supports clinicians in diagnoses and treatment validation. On the other hand, simply the immense amount of data overwhelms the abilities of trained experts as well as the capacities of the health systems. Automated procedures may present a remedy to this dilemma.

In this wonderful and enjoyable book, Dagmar Kainmüller addresses automatic segmentation of medical images and contributes significantly to the art of segmentation. On the basis of statistical shape models, Dagmar develops new tools which overcome many of the drawbacks of current state alternatives. Remarkable and price-awarded contributions of Dagmar are the introduction of ODDS (Omnidirectional Displacements for Deformable Surfaces) and mesh coupling for multi-object segmentation, to name just two. Most importantly, Dagmar presents solutions to real-life applications such as liver segmentation and hip and knee segmentation.

Many more treasures lie in Dagmar's book and certainly deserve an extended acknowledgement, but I like to keep the reader curious and conclude with another quote of William Shakespeare from "King Henry the Fifth" (Act III, Scene II):

"Men of few words are the best men."

Prof. Dr. Jan Modersitzki

Institute of Mathematics and Image Computing, University of Lübeck, Germany
Fraunhofer MEVIS Project Group Image Registration, Lübeck

Acknowledgments

I wish to thank Jan Modersitzki for supervising my thesis. Jan, I am extremely grateful for everything you taught me! I wish to thank Stefan Zachow and Hans Lamecker: Thank you with all my heart for your help and advice, for your trust in my work, and for your friendship. I wish to thank Hans-Christian Hege for his support throughout my thesis. Many thanks to Matthias Bindernagel and Judith Berger for their great work. Many thanks to Thomas Lange for an awesome collaboration. A special thanks to Heiko Ramm: You are the best office-mate in the world! Tons of thanks to Andrea Kratz, Britta Weber, Conni Auer, Daniel Baum, and to the Medical Planning group for their help and support and for the incredibly nice time I had working at ZIB.

It has been a pleasure working with you. Thank you!

Abstract

Segmentation of anatomical structures in medical image data is an essential task in clinical practice. Clinical applications that call for image segmentation include virtual surgery planning, therapy planning, diagnosis, and patient monitoring. This thesis contributes methods for accurate fully automatic segmentation of certain anatomical structures in 3D medical image data. It follows the Deformable Model approach for segmentation, and makes use of Statistical Shape Models.

The core methodological contribution of this thesis is a novel deformation model for triangle meshes that overcomes limitations of state-of-the-art approaches. This deformation model allows for accurate segmentation of tip- and ridge-shaped features of anatomical structures. Concerning methodology, a second focus of this work lies on accurate multi-object segmentation.

As for practical contributions, this thesis proposes application-specific segmentation pipelines for a range of anatomical structures, together with thorough evaluations of segmentation accuracy on clinical image data. These fully automatic pipelines allow for highly accurate segmentation as compared to related work. E.g., the pipeline proposed for segmentation of the liver in contrast-enhanced CT is, as at June 2014, the most accurate among all competing fully automatic approaches on benchmark image data.

Contents

Chapter 1

Introduction

Contents

This thesis contributes methods for *accurate fully automatic segmentation* of *certain anatomical structures* in *3D medical image data*. Section 1.1 introduces basic terms and motivates the objective of this thesis. Section 1.2 describes the methodological and practical context in which this thesis was realized, lists the specific contributions of this work, and distinguishes related topics that are not discussed. Section 1.3 gives an overview of the structure of this thesis.

1.1 Motivation

Section 1.1.1 introduces what 3D medical image data is and what a segmentation of an anatomical structure is, and describes the purpose and benefit of such segmentations. Section 1.1.2 describes the task of *manual segmentation* and motivates

1

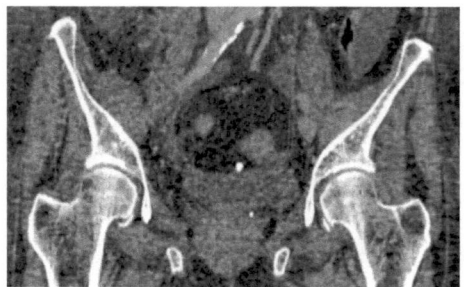 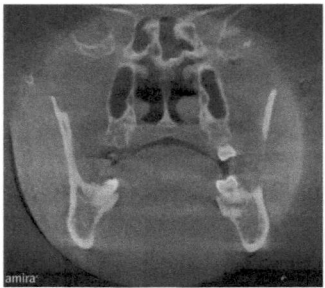

Figure 1.1 2d slices of exemplary image data. Left: Coronal slice of CT of the pelvic bones. Right: Coronal slice of cone-beam CT of the mandible. Note that we employed the ZIBAmira software (see Stalling et al. (2005) and amira.zib.de) to create most of the figures shown in this thesis.

the need for *automation*. Section 1.1.3 discusses the *correctness* and *reproducibility* of manual segmentations, as well as the *accuracy* of both manual and automated segmentations.

1.1.1   Segmentation of Medical Image Data

Three-dimensional, *tomographic* medical imaging has become an essential tool in diagnosis and intervention planning in clinical practice (see e.g. Sonka and Fitzpatrick (2000); Pham et al. (2000)). Tomographic images can be acquired with various *imaging modalities*. Image intensities capture some physical property of the tissue at the respective location. The particular property depends on the specific imaging modality. E.g., *X-ray computed tomography* (CT) measures the radiodensity of tissue, while *magnetic resonance imaging* (MRI) can measure (among other properties) the density of hydrogen nuclei. For more details on image acquisition, see e.g. Udupa and Herman (2000); Handels (2009). 3d medical images are *voxel images*, where a voxel is the 3d equivalent of a pixel in 2d. A voxel image can be seen as a stack of pixel images. For straightforward visualization of a 3d medical image, 2d planar sub-images can be displayed, where the image plane is spanned by two axes of a reference coordinate frame. Such sub-images are called *slices* of the image. Reference coordinate axes are usually aligned with *body axes* of the imaged person, where the first axis points from right to left side of the body, the second axis points from front to back, and the third axis points from feet to head. Image slices parallel to the (x_1, x_2)-plane are called *axial* or *transversal* slices, parallel to the (x_1, x_3)-plane *coronal* slices, and parallel to the (x_2, x_3)-plane *sagittal* slices. Figure 1.1 shows exemplary slices of 3d medical images.

For many applications in clinical practice, organs or other anatomical structures

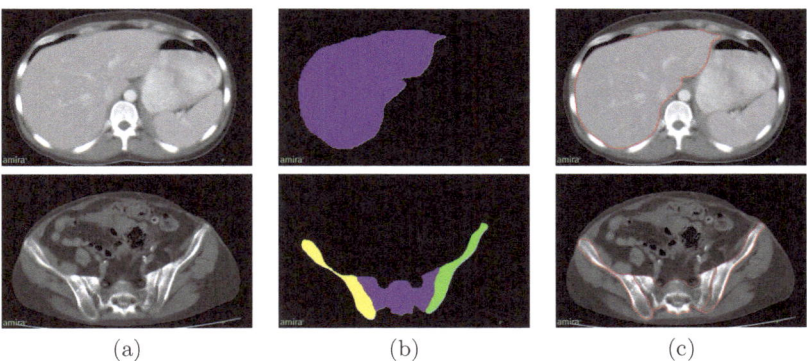

(a)   (b)   (c)

Figure 1.2 Slices of exemplary image data with segmentations. (a) Top: Axial slice of contrast-enhanced liver CT. Bottom: Axial slice of pelvis CT. (b) Respective segmentations as color images of the same slice. Liver: *Binary* segmentation. Pelvis: Individual colors for left hip bone, sacrum, and right hip bone. (c) Delineations (red) of the respective structures overlaid on image slices.

of interest need to be *delineated* or *crayoned* in medical images. This is referred to as *segmentation* (see e.g. Pham et al. (2000)). Figure 1.2 shows segmentations of exemplary anatomical structures in 2d slices of 3d medical image data. Segmentations of anatomical structures in tomographic medical image data provide 3d models capturing the respective structures' spatial extension within an individual patient's body. Such 3d *geometrical* models of patient specific anatomy, potentially enriched with material properties, allow physicians to perform computer simulations of planned surgical or therapeutical interventions, and analyze the respective outcome. In case of unsatisfactory (simulated) outcome, the physician can decide to alter the surgical or therapeutical procedure, until a plan with satisfactory outcome is found. This procedure is referred to as *virtual surgery- and therapy planning*. Procedures that can be planned in such a way are e.g. liver resection (cf. Section 7.1) and total hip or knee replacement (cf. Chapter 8). Further applications of image segmentation in clinical practice include diagnosis, monitoring the progression of a disease, and monitoring the response to a therapy. Applications are also in clinical research, where segmentations are e.g. of use for the study of anatomical structure (Pham et al., 2000). For a more thorough discussion of the role of three-dimensional medical image segmentation in the clinic, see e.g. Udupa and Herman (2000); Handels (2009).

1.1.2 Automation of the Segmentation Task

Segmentations can be generated manually: Therefore, an expert has to assign a label to each voxel of an image. Practically, each structure of interest has to be "crayoned" in some label color in each slice of the image stack. Consistency of labelings in neighboring slices has to be checked by examining orthogonal slices. Overall, manual segmentation is an extremely tedious and time-consuming task, which limits the feasibility of intervention planning in clinical practice given the large number of cases that need to be treated. For example, several hours are necessary for manual segmentation of the pelvic region for radiotherapy planning (Pekar et al., 2004) and likewise for manual segmentation of knee bones and cartilage for osteoarthritis treatment (Shim et al., 2009), and manual segmentation of the liver for resection planning can take up to three hours (Beichel et al., 2001). This motivates the need for automated segmentation methods.

In this thesis, we aim at *fully automatic* methods for segmentation. We use the term "fully automatic" to refer to algorithms that do not require any manual interaction during their run-time, and also do not require any manual choice or tuning of parameters for individual images beforehand. Thereby we follow the common usage of the term in the field of automated segmentation, where the predicate "fully automatic" distinguishes a method from "semi-automatic" and "interactive" methods (see e.g. van Ginneken et al. (2007)).

Any clinical application requires the physician to *validate* such fully automatic segmentations, and correct them with manual interaction if necessary. The goal of fully automatic segmentation is to reduce the overall manual effort. Therefore, an automatic segmentation has to be either "good enough" to not require any manual correction (in this case eliminating any manual effort), or "good enough" to only require "small" manual corrections, i.e. corrections that can be performed fast and easy compared to the manual effort that would be required without the fully automatic segmentation.

Concerning the use of fully automatic segmentation algorithms in clinical practice, Heimann and Meinzer (2009) write (in the context of statistical shape models for segmentation):

> "Although we do not believe that a fully automatic procedure will be able to segment 100% of images successfully in the foreseeable future, efforts to lower the number of complete failures have to be intensified. This will increase acceptance [...] in the clinic and lead to wider application and benefit."

and also in the context of liver segmentation (Heimann et al., 2009):

> "Still, the obvious advantage of most interactive systems [...] is the

complete user control over the result. This user control is required for clinical applications as long as automatic methods still fail on certain image data."

In other words, although fully automatic methods exist that produce satisfactory segmentations on selected image data, overall, manual corrections are still necessary to an extent that prohibits a widespread use of such methods in clinical practice. This is despite decades of research on digital image processing, where the medical field is one among many areas of application (see e.g. Gonzalez (2009)). We will discuss particularities of medical images that renders automated segmentation especially difficult in Section 1.2.1.

1.1.3 Segmentation Accuracy

An objective assessment of the accuracy of a segmentation implies knowledge of the "correct" segmentation, i.e. knowledge of the actual location of the anatomical structure within the patient. Concerning medical image data, the correct segmentation of an anatomical structure is usually not known (except for *phantom-* or cadaver studies). Instead, a manual segmentation performed by an expert usually serves as "ground truth". However, manual segmentations performed by multiple experts – and also by one expert at different times – typically yield different results, a phenomenon which is referred to as *inter- and intra-observer variability.*

The ability to *compare* two segmentations by means of quantitative distance- or dissimilarity measures is vital to (1) quantify inter- and intra-observer variability, and (2) quantify the *accuracy* of an automatic segmentation or a manual segmentation performed by an inexperienced person *w.r.t. the ground truth.* Note that quantified inter- and intra-observer variability indicate how *reproducible* a structure can be segmented – in general, this is not a measure of correctness.

We describe a set of "standard" quantitative dissimilarity measures in Section 6.1.

1.2 Scope of this Thesis

In this thesis, we follow the *Deformable Model* approach for automated segmentation, as introduced in Section 1.2.1. The selection of applications as presented in this thesis is motivated in Section 1.2.2. The contributions of this thesis – both methodological and practical – are outlined in Section 1.2.3. Section 1.2.4 discriminates related topics not discussed in this thesis.

1.2.1 Automated Segmentation with Deformable Models

As Tönnies (2012) puts it, substantially unisonous with others (e.g. (Sonka and Fitzpatrick, 2000; Birkfellner, 2011)),

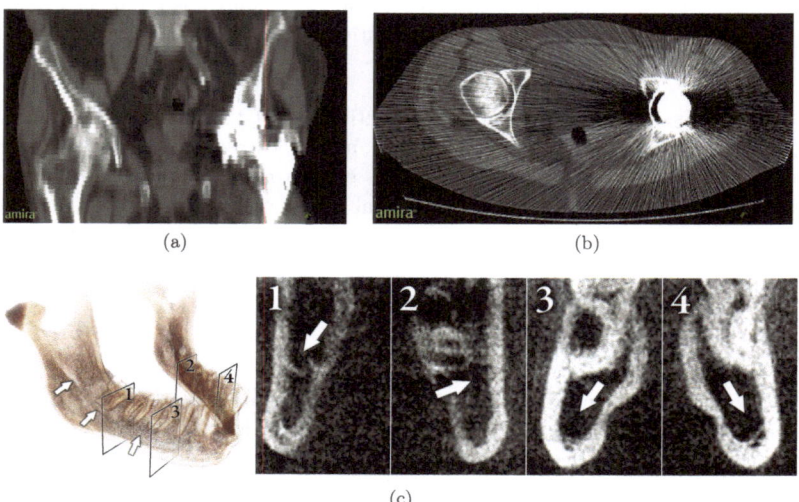

(a)   (b)

(c)

Figure 1.3 Exemplary imaging deficiencies and low contrast. (a) Partial volume effect obscures the exact location of the hip joint gap in hip CT. (b) *Metal artifacts*: Streaks originating from metallic hip joint implant overshadow hip CT slice. (c) *Volume rendering* (left) and slices (right) showing mandible in cone-beam CT. The mandibular nerve, indicated by arrows, exhibits low contrast to surrounding bone marrow.

"medical images are different from other pictures."

Automated segmentation of medical images is difficult due to imaging deficiencies such as noise, artifacts, and *partial volume effects* i.e. spatial aliasing stemming from the discrete nature of three-dimensional medical images (see e.g. Kalender (2011)). Figure 1.3a,b shows exemplary image slices. Apart from these imaging deficiencies, a specific anatomical structure can only be depicted as "contrasted" as permitted by the physical properties of the respective image acquisition modality – for example, an adjacent anatomical structure may appear with similar intensity characteristics and hence may be hard to distinguish from the sought structure. Figure 1.3c shows an exemplary situation. Hence, for many applications, automated segmentations generated with "low-level" techniques such as thresholding and region growing (see e.g. Handels (2009), Chapters 5.1 and 5.3) are nowhere near being "good" representations of the sought anatomical structure. Thresholding most probably produces segmentations with false topology in the presence of noise, while region growing from a manually defined seed point within the sought structure will often "leak" into neighboring tissue. Figure 1.4 shows exemplary results of thresholding and region growing.

A priori knowledge about location, size, *shape* and *appearance* of a sought structure "facilitates" its segmentation in the presence of imaging and modality deficiencies. We employ the term *shape* to refer to a geometrical description of an object – or equally its external boundary – that is invariant to pose changes. We employ the term *appearance* to refer to intensity characteristics an object exhibits when imaged with a specific imaging modality. Note that we use both terms in a "natural language" sense, i.e. without precise mathematical definition, which is a common usage in the context of deformable models (see e.g. Cootes et al. (1995)).

Deformable Models (McInerney and Terzopoulos, 1996; Xu et al., 2000; Montagnat et al., 2001) are a segmentation technique that provide a machinery to exploit such a-priori knowledge, and have been shown capable of coping with imaging deficiencies. The basic idea is to deform a given *template shape* in such a way that the deformed shape provides a geometric representation of the corresponding structure in the image. A design choice in the deformable model approach is the *shape representation* which is used to describe the geometry of the template shape. Another ingredient is the *deformation model*, which captures knowledge about the target structure's shape variability and thereby restricts the deformation of the template shape to only yield plausible shapes. Furthermore, an *appearance model* captures knowledge about how the target structure is expected to appear in the image. These three ingredients allow the image segmentation problem to be encoded as the problem of optimizing an *objective*. The search for an optimal deformation w.r.t. this objective is performed with some *search algorithm*.

While *generic* knowledge about a shape's properties such as its topology or smoothness yields deformation models that are beneficial for *interactive* (i.e. manually guided) segmentation, *automated* Deformable Model approaches call for *specific*, i.e. "application-tailored" or "learned" models (McInerney and Terzopoulos, 1996).

Statistical shape models (SSMs) as introduced by Cootes et al. (1995) are particular deformation models which are highly application specific (McInerney and Terzopoulos, 1996). The basic idea is to learn the "natural" shape variability of a particular structure from a training set of individual shapes, yielding a model which is highly specific for this structure. That means the model can describe – to some accuracy – the shape of any individual instance of the particular structure, but cannot describe "anything else".

SSMs are widely used and well-established for modeling *anatomical structures* with specific (complex, distinguished) shapes, as e.g. the pelvic bones and the liver, and segmenting the respective structures in medical image data (Heimann and Meinzer, 2009). The work presented in this thesis employs and builds upon SSMs for segmentation.

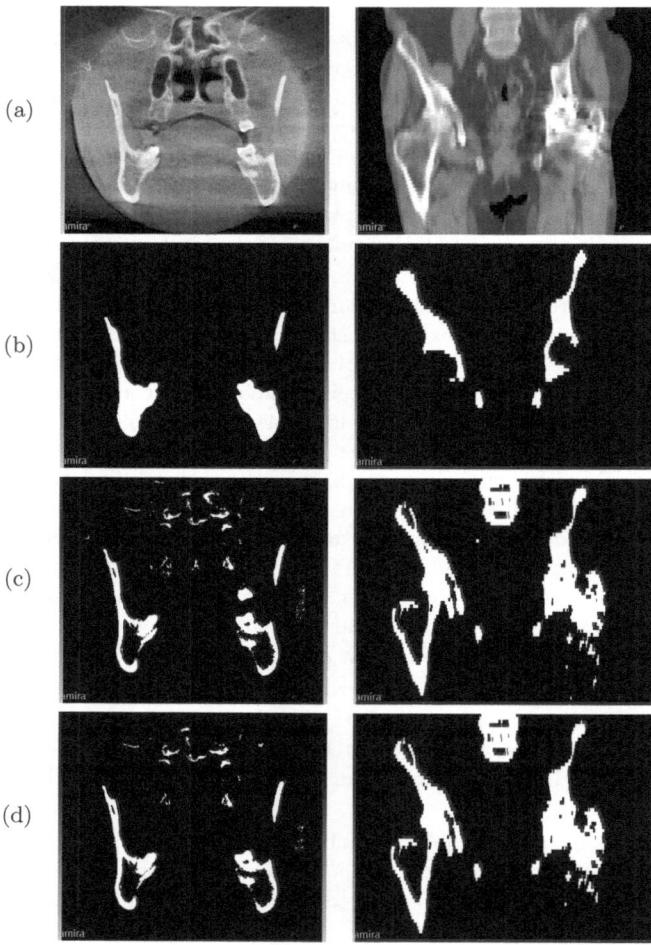

Figure 1.4 Exemplary results of thresholding and region growing. Left column: Mandibular bone in cone-beam CT. Right column: Pelvis in CT. (a) Slices of image data. (b) Respective ground truth segmentations. (c) Intensity thresholding results. (c) 3d region growing from seed point within mandibular bone and pelvis, respectively, with same intensity thresholds as in (c). Region growing captures only slightly less adjacent tissue and noise than thresholding.

1.2.2 Selection of Anatomical Structures and Imaging Modalities

This thesis builds upon the work of Lamecker et al. (2002, 2004a, 2006b) who have presented SSMs for a series of anatomical structures including the liver, individual bones, and substructures of bones, and showed their potential for automated segmentation. These structures have in common the characteristics of exhibiting a distinguished shape and constant topology among individual subjects. Also, they typically occupy a relatively large part of the image domain, which makes them easy to *detect*. This thesis proposes methods and algorithms for fully automatic and accurate segmentation of these structures. Furthermore it deals with adjacent structures less suitable for modeling via SSMs, namely soft-tissue structures of the musculoskeletal system (cartilage and muscles).

As for the selection of imaging modalities, this thesis deals with "standard" modalities indicated for the respective planning tasks: Joint replacement surgery is typically planned based on CT data (cf. Section 8.1), osteoarthritis treatment is performed on the basis of MRI data (cf. Section 8.2), cone-beam CT data is increasingly popular for dental applications (cf. Section 7.3), and contrast-enhanced CT data serves for liver resection planning (cf. Section 7.1). The only non-standard imaging modality dealt with is CT (instead of MRI) for the task of muscle segmentation, as motivated in Section 10.2.

1.2.3 Contribution

This work contributes both methodological and practical advances towards accurate and fully automatic segmentation of selected anatomical structures in medical image data. As for methodological contributions, it proposes

- A novel *deformation model* for accurate segmentation of highly curved regions on anatomical structures (Chapter 4), which is able to improve segmentation accuracy as compared to conventional approaches, as shown in an extensive evaluation on clinical data (Chapter 9);

- A novel *mesh-coupling* algorithm allowing for accurate and consistent (non-overlapping) multi-object segmentation of arbitrary adjacent structures via simultaneous deformations (Section 3.5). Again, an evaluation on clinical data reveals improved accuracy as compared to conventional approaches (Section 8.1).

As for practical application-specific engineering, this work contributes fully automatic segmentation pipelines for selected anatomical structures that yield very accurate results as compared to other automatic approaches, namely

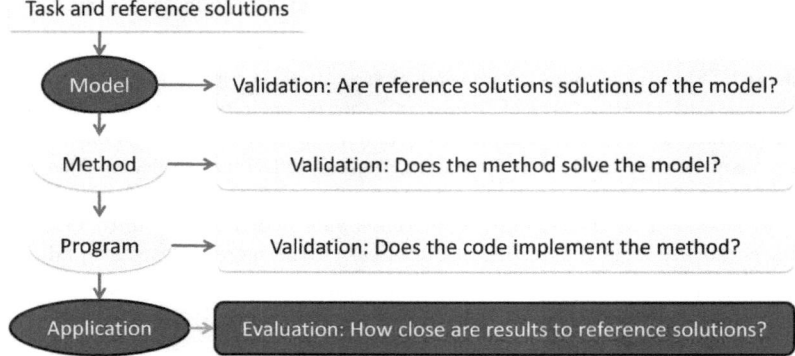

Figure 1.5 Problem solving pipeline. First, the problem or task is formulated by means of a *model*. Second, the model is implemented by means of a *method*. Third, the method is implemented by means of a computer *program*. Finally, the program is *applied* to input data to generate *results*. The steps we focus on in this thesis are highlighted in red/gray.

- As at June 2014 the *most accurate system for liver segmentation in contrast-enhanced CT* according to a comparative evaluation on benchmark image data (Section 7.1);

- A system for *knee joint segmentation in MRI* (bones and cartilage) which as at June 2014 is the 2nd most accurate method according to a comparative evaluation on benchmark image data (Section 8.2)

- A system for *segmentation of the pelvic bones in CT* (Section 7.2) where a step-by-step evaluation reveals the benefit of each step in the pipeline. Though this system outperforms previous approaches in terms of average accuracy measures, results are not directly comparable since no public pool of benchmark pelvic CTs is available for evaluation.

Furthermore, this work presents fully automatic segmentation pipelines for *novel applications*, namely

- The first reported and quantitatively evaluated systems for fully automatic segmentation of the *mandibular bone and nerves in cone-beam CT* (Section 7.3), and *individual muscles in CT* (Section 10.2).

1.2.4   Topics Not Discussed

To point out and put into context the focus of this thesis, Figure 1.5 sketches a generic problem solving pipeline. The *task* tackled in this thesis is segmentation of medical images. Reference solutions are given by means of gold-standard segmentations. Typically, the task is *modeled* as an *optimization problem*. A method which solves an optimization problem is called *solver*. The method is implemented as a computer program, which can then be applied to solve the task on input data. Each step in this pipeline is subject to validation.

This thesis focuses on modeling, applications, and thorough quantitative and qualitative evaluation of segmentation results in terms of accuracy. The quality of a model, i.e. how well a model represents the task, is subject to thorough analysis – a solution of a model does not necessarily lead to accurate segmentation results. However, model validation is not a topic of this thesis. We tackle optimization problems with off-the-shelf methods, namely discrete or numerical solvers. Their theoretical properties (e.g. concerning convergence and optimality of results) are denoted as far as cited in the literature – yet "good" theoretical properties of solvers are not a major concern of this thesis. One minor concern however is a "manageable" *run-time* of the program code that implements methods.

With the goal of making intervention planning feasible for the large numbers of cases that are treated in clinical practice, this work aims at reducing the manual effort necessary for image segmentation. However, evaluations on clinical data as presented in this work target the *accuracy* of automatic segmentations. The actual *time savings* the presented methods provide in clinical practice are not assessed quantitatively. In this regard another important question is how accurate a segmentation "has to be" to be of use in a particular application. This is also not discussed in this thesis. Instead, we aim at "the highest achievable" accuracy for all applications.

This work follows the *Deformable model* approach (cf. Sec. 1.2.1) for *fully automatic* segmentation of *selected* anatomical structures (cf. Sec. 1.2.2) from 3d medical image data. *Triangle meshes* serve as shape representation, as motivated in Chapter 3. Alternative shape representations, as e.g. *level sets* (Malladi et al., 1995), are not discussed. *Heuristic* (i.e. manually tailored) appearance models are established and employed in this thesis, as motivated in Chapter 3. *Statistical* models of appearance, as e.g. *Active Appearance Models* (Cootes et al., 2001), are not discussed.

Computational methods that facilitate manual segmentation or allow for efficient *interactive correction* of automated segmentations (see e.g. Schenk et al. (2000); Hamarneh et al. (2005); Liu et al. (2008)) also serve the purpose of reducing manual effort and are vital components of planning systems – yet they are not discussed in this thesis.

1.3   Structure of this Thesis

The remainder of this thesis is organized as follows:

Four Chapters (2-5) on methodologies form the first Part. Chapter 2 introduces basic terms and their notation as well as basic definitions which we employ throughout the following chapters. Chapters 3 and 4 provide a model kit of methods that can be assembled to form fully automatic segmentation pipelines for SSM-modellable structures. Chapter 5 proposes options for extending this kit to "nearby" structures.

In particular, Chapter 3 introduces *deformable meshes* for segmentation. It describes all components necessary to form fully automatic pipelines for accurate segmentation of structures that can be modeled by means of *statistical shape models*. The respective pipelines are robust w.r.t. noise and artifacts in medical image data, and can handle low contrast to adjacent structures.

The methods described in Chapter 3 are "conventional" in the way they search for *features* that an anatomical structure is expected to exhibit the image data: Image appearance is analyzed in one-dimensional subsets of the image data for each vertex of a deformable mesh. This conventional search has certain disadvantages, leading to inaccurate segmentation results for highly curved regions of anatomical structures. Chapter 4 proposes a novel search strategy and *deformation model* to overcome this limitation.

Not every anatomical structure can be modeled by an SSM. The success of an SSM depends on the specific kind of variability of the modeled structure. Anatomical structures that appear with varying topology, or non-linear shape variations like articulation, are not suitable for modeling with SSMs. Chapter 5 proposes options for extending segmentation pipelines to structures that are not suitable for modeling with SSMs, yet have some specific spatial relation to SSM-modellable structures. The basic idea is to *extrapolate* such structures from nearby structures that *can* be modeled and segmented with the SSM-based framework, and refine the resulting extrapolation via *atlas-based segmentation*.

Part II of this work, formed by Chapters 6-10, presents a series of clinically relevant applications of the methods described in Part I, together with thorough quantitative evaluations of segmentation results. Chapter 6 introduces the basics of quantitative evaluation and parameter settings.

Chapter 7 describes three applications and respective application-specific combinations of methods (i.e. *pipelines*) for *single-object segmentation*. These pipelines are assembled from methods described in Sections 3.1-3.4. The segmentation accuracy of each pipeline is evaluated on pools of clinical data. The system for liver segmentation in contrast enhanced CT as presented in Section 7.1 was awarded the first prize in an international competition on accurate automatic liver segmentation. However, inaccuracies can be observed due to the "conventional" search for image features, motivating the method contributed in Chapter 4. Section 7.2 describes a system for segmentation of the pelvic bones in CT which is formed by a combination of steps that suits the particular characteristics of the application and is different from the liver segmentation system. The accuracy of results is assessed in a step-by-step evaluation on 50 clinical CTs. Again, inaccuracies due to "conventional" search for appearance match can be observed, as well as inaccuracies in joint regions, which exhibit low contrast to the adjacent femoral bone, motivating the methods contributed in Chapter 4 and Section 3.5. Section 7.3 proposes a system for a novel application, namely segmentation of the mandibular bone and nerves in cone-beam CT. A step-by-step evaluation is performed on a large set of clinical data. The sub-system for segmenting alveolar nerve channels provides an exemplary pipeline for segmentation of SSM-modellable *line-like* structures.

Chapter 8 describes two applications that call for *multi-object segmentation* methods as presented in Section 3.5. Section 8.1 applies a mesh-coupling approach that enables multi-object segmentation of arbitrary adjacent structures for segmentation of the hip joint in CT. Segmentation accuracy is evaluated on a pool of clinical data. Section 8.2 presents a complex pipeline employing multi-object segmentation methods in combination with application specific engineering for accurate segmentation of knee bones and cartilage in MRI data. This pipeline was awarded the second prize in an international competition on accurate knee segmentation.

Chapter 9 presents an extensive evaluation of our novel deformation model for highly curved structures as presented in Chapter 4 on large sets of clinical data, with the coronoid process of the mandibular bone and the acetabular rim of the hip bone serving as exemplary highly curved anatomical structures.

Chapter 10 presents two applications employing extended pipelines as proposed in Chapter 5. Section 10.1 evaluates different options for extrapolating nearby anatomical landmarks from surface meshes of the pelvic bones. Section 10.2 proposes a pipeline for fully automatic segmentation of individual muscles of the pelvic region in CT and presents preliminary results on clinical data.

The thesis concludes with a summary of its contributions, a discussion of its impact both on the field of medical image analysis and on clinical practice, and considerations on future work.

Part I

The Segmentation Framework

Chapter 2

Basic Terms and Notation

Contents

This chapter introduces basic terms and notations which we use throughout the following chapters.

Lists $A := \{a_i\}_{i=1}^{n_A}$ appear in different contexts throughout this thesis. We denote the index set of a list A as $N_A := \{1, \ldots, n_A\}$. We shortly denote that an element a appears in a list, i.e. $\exists i \in N_A : a = a_i$, as $a \in A$. We denote list B appended to list A as $A \cup B$. Whenever it adds to clarity of notation, we denote a function f whose domain is the index set N_A sloppily also as a function whose "domain" is A.

For a finite list $A := \{a_i\}_{i=1}^{n_A}$ and a function $y : N_A \to \mathbf{R}$, in case y is not guaranteed to have a unique minimizer, with $i := \min\left\{\operatorname{argmin}_{j \in N_A} \{y(j)\}\right\}$ we denote the respective element of the list, a_i, sloppily as $\operatorname{argmini}_{a \in A} \{y(a)\}$. We define argmaxi analogously. This definition is convenient in case an objective y is not guaranteed to have a unique minimizer, yet we need to decide for one single minimizer without having a sophisticated criterion for this decision at hand. In this case argmini helps us denote one minimizer distinguished by means of a straightforward criterion, namely the minimizer with minimal index in the list.

2.1   Images, Segmentations, and Surface Meshes

2.1.1   Three-dimensional Medical Images

Three-dimensional medical images are *digital* images (see e.g. Handels (2009)). Intensities are acquired only at a finite set of locations $X \subset \mathbf{R}^3$. A location $x \in X$ together with an intensity $I(x) \in \mathbf{R}$ as acquired for this location is called *voxel* (see e.g. Smith (1995)). Locations X are organized as a regular grid. One commonly used grid is

$$X = \left\{ \mathbf{x} \in \mathbf{R}^3 : \forall c \in \{1,2,3\} \, \exists i \in \{0, \ldots, n_c - 1\} : x_c = x_c^{(0)} + \delta_c \cdot i \right\} , \quad (2.1)$$

where $\delta_c \in \mathbf{R}^+$ denotes the distance between locations for which intensities are acquired in direction of coordinate c, and $n_c \in \mathbf{N}^+$ denotes the number of such locations. The location $\mathbf{x}^{(0)} \in \mathbf{R}^3$ is the *origin* of the image. Due to the details of image acquisition, there is often one distinguished plane with isotropic resolution, i.e. $\delta_i = \delta_j$ for $i, j \in \{1, 2, 3\}$, $i \neq j$, while resolution is lower in direction orthogonal to this plane, i.e. $\delta_k \neq \delta_i$ for $k \in \{1, 2, 3\} \setminus \{i, j\}$. In consequence, grids are in general anisotropic.

The convex hull of X yields an interval $\Omega \subset \mathbf{R}^3$. Image intensities for locations $y \in \Omega \backslash X$ can be derived by means of *interpolation*. One widely used interpolation method assigns to y a convex combination of intensities as given at the eight closest locations contained in X, computed via subsequent linear interpolations in each coordinate direction. This approach is called *trilinear interpolation*. Another approach assigns to y the intensity given at the closest location $x \in X$. This is referred to as *nearest neighbor interpolation*. For more details on grids and interpolation see e.g. Modersitzki (2004).

A digital image together with an interpolation method yields a function

$$I : \Omega \to \mathbf{R} .$$

In the following such a function I derived from a grid (2.1) and trilinear interpolation is our mathematical model for a three-dimensional (3d) medical image. It assigns intensities (gray-values) to an interval $\Omega \subset \mathbf{R}^3$ that contains (parts of) the space that a human body occupies within some reference coordinate system.

2.1.2   Segmentations of Three-dimensional Medical Images

A *segmentation* of an image is a finite partition of the image. It assigns a region-ID or *label* to each location in the image domain, $\Omega \to \{0, 1, \ldots, n\}$, with n the number of labels. The "classical" definition of a segmentation requires locations equipped with the same label to be *similar* in terms of image characteristics (see e.g. Pham et al. (2000)). This definition is suitable to characterize image partitions as generated by automatic, low-level segmentation methods. In this thesis, we use the term "segmentation" to also refer to image partitions that do not necessarily comply with this definition. Particularly, we call a partition "segmentation" if its intended purpose is to assign the same label to locations that share *semantic information* such as *belonging* to the same organ or other anatomical structure (see e.g. Heimann (2003)).

A *binary* segmentation partitions the image domain into two subsets called foreground (0) and background (1): $\Omega \to \{0, 1\}$. A segmentation of a voxel image that assigns a unique label to each voxel is called *hard segmentation* (see e.g. Pham et al. (2000)). As opposed to *soft* or *probabilistic* segmentations, hard segmentations do not allow for sub-voxel accuracy of segmentations, nor do they capture information about segmentation uncertainty. Although sub-voxel accuracy and segmentation uncertainty are highly interesting and valuable concepts, this thesis does not pursue this path of research, but sticks to conventional, hard segmentations. In the following, we use the terms *hard segmentation*, *voxel segmentation* and *segmentation* synonymously.

A hard segmentation implies a voxel labeling $X \to \{0, 1, ..., n_R\}$. A labeling of Ω is commonly derived from a voxel labeling by means of nearest neighbor interpolation. In this sense, a *segmentation of an anatomical structure in a voxel image* is a binary partition that assigns "foreground" (label 1) to each voxel of which $> 50\%$ volume belong to the anatomical structure, and "background" (label 0) to all others. A segmentation of multiple anatomical structures in a voxel image assigns label l to each voxel of which $> 50\%$ volume belong to structure l, and label 0 to all others.

Note that the term *segmentation* refers to both the task and the result of partitioning, i.e. *segmenting*, an image.

2.1.3 Triangle Surface Meshes

Following Schneider and Eberly (2003, Chapter 9.3), we call *triangle mesh* a finite collection of *vertices*, *edges*, and *triangles* that satisfies certain conditions as described in the following. A vertex is a point in space, $v \in \mathbf{R}^3$. We refer to the list of vertices of a triangle mesh as $V := \{v_i \in \mathbf{R}^3\}_{i=1}^{n_V}$, where n_V is the number of vertices. We refer to the index set of V as $N_V := \{1, \ldots, n_V\}$. We call the concatenation of all vertices of a mesh *shape vector*, $\mathbf{v} := (v_1^T, ..., v_{n_V}^T)^T \in \mathbf{R}^{3n_V}$. An edge is a line segment that connects two different vertices. An edge is identified by a tuple formed by the respective vertices' indices, $(j, k), j, k \in N_V, j \neq k$. We refer to the list of edges $E \subset N_V \times N_V$ as $E := \{(j, k)_i\}_{i=1}^{n_E}$, where n_E is the number of edges. A triangle is the convex hull of three different vertices. A triangle is identified by a triple formed by the respective vertices' indices, $(j, k, l), j, k, l \in N_V, j \neq k \neq l \neq j$. We refer to the list of triangles (faces) $F \subset N_V \times N_V \times N_V$ as $F := \{(j, k, l)_i\}_{i=1}^{n_F}$, where n_F is the number of triangles.

The conditions that the collection (V, E, F) has to satisfy to be a *mesh* are that each vertex belongs to at least one edge, and each edge belongs to at least one triangle. If any two triangles of a mesh are connected by a path that leads from triangle to triangle over *shared edges* (i.e. edges that belong to multiple triangles), the mesh is called *connected*. A connected mesh is called *manifold* if each edge is shared by at most two triangles. A manifold mesh is called *orientable* if triangle triples can be ordered such that each edge (j, k) of the mesh that is shared by two triangles appears in order j, k in one triple and in reverse order, k, j, in the other triple. Informally speaking this implies that the mesh has "inside" and "outside". A mesh is called *closed* if any edge is shared by exactly two triangles. Closedness implies the mesh to be manifold.

If not noted otherwise, in the following we use the term *triangle surface mesh* or just *surface mesh* to refer to a closed orientable mesh $M := (V, E, F)$. Note that we deviate from Schneider and Eberly (2003) only by not considering *self-intersections* of meshes in the above definitions. A mesh has self-intersections if it has faces, edges or vertices that interpenetrate each other. Self-intersecting meshes appear throughout this thesis – we call them *meshes* for ease of terminology, and discuss the impact of self-intersections if it is important in a certain situation.

Any point that *lies on a surface mesh* $M := (V, E, F)$ is defined by means of a triangle $t := (j, k, l) \in F$ it lies on and by its *barycentric coordinates* (α, β, γ) on this triangle (see e.g. Coxeter (1969)). Its location $x \in \mathbf{R}^3$ is

$$x = \alpha v_j + \beta v_k + \gamma v_l \text{ with } \alpha, \beta, \gamma \in [0, 1], \alpha + \beta + \gamma = 1 .$$

In other words, as $\gamma = 1 - \alpha - \beta$, every point on M is described by a triple (t, α, β) with $t \in F$ and $\alpha, \beta \in [0, 1]$. In reverse every such triple describes a point on M. In case of self-intersections of the mesh, there are point positions $x \in \mathbf{R}^3$ on M that are described by multiple triples, yet one triple always describes exactly one point. For a mesh $M := (V, E, F)$ we denote

$$\mathcal{F} := \{(t, \alpha, \beta) : t \in F, \, \alpha, \beta \in [0, 1]\} .$$

We denote the respective set of point position in $\mathbf{R}^3$ as

$$\mathfrak{M} := \{\alpha v_j + \beta v_k + \gamma v_l : ((j, k, l), \alpha, \beta) \in \mathcal{F}, \gamma := 1 - \alpha - \beta\} .$$

There is a bijection between $\mathcal{F}$ and $\mathfrak{M}$ if and only if the respective mesh M is free of self-intersections.

The surface normal at vertex v of a surface mesh, denoted as n_v, can be estimated via the set of edges of the mesh that contain the vertex. Similarly, the principal

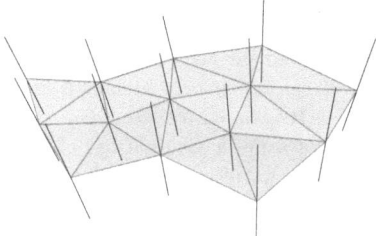

Figure 2.1 Detail of a triangular surface mesh (red/gray). Vertex normals indicated by black line segments.

curvatures at a vertex can be estimated from a vertex neighborhood (Hildebandt et al., 2005). Figure 2.1 shows a detail of a triangle mesh.

2.1.4 From Segmentations to Surface Meshes and Back

A binary label image can be converted into a surface mesh that represents the boundary of the volume labeled "foreground" by means of the *Marching Cubes* or *Generalized Marching Cubes* algorithms (Lorensen and Cline, 1987; Hege et al., 1997). In reverse, a surface mesh can be converted into a binary label image by means of *Scan Conversion* (Kaufman, 1987). With the same algorithms, multi-label images can be converted into meshes (which are in general non-manifold), and vice-versa.

2.2   Deformable Surface Meshes

Deformable surface meshes are surface meshes which are deformed by means of vertex re-locations with the goal of yielding a geometric representation of a target structure sought in an image. For each vertex of a deformable surface mesh, a set of *candidate locations*, i.e. potential new vertex positions, are tested for *appearance match*. This is done by comparing actual image appearance at candidate locations to an *appearance model*. Therefore, at each candidate location, image information is assessed in a certain neighborhood, as required for comparison with the particular appearance model. Given this analysis of appearance match, a new shape is computed by *displacing* the vertices of the mesh to suitable locations, following a trade-off between appearance match and anatomically plausible deformation.

2.2.1 Displacement Fields and Sets of Candidate Displacements

A surface mesh is *deformed* by moving each of its vertices to a new position, i.e. by *displacing* each vertex. A *displacement field* assigns a *displacement* $s \in S$ to each vertex of a mesh, where $S \subset \mathbf{R}^3$ is a set of *candidate displacements*. Assigning a displacement to each vertex of a mesh yields a list of displacements $\{d_i \in S\}_{i=1}^{n_V}$. If it adds to clarity of notation, we also sloppily denote a displacement field as a "function" on the list of vertices V, $d : V \to S, v_i \mapsto d(v_i) := d_i$. We denote an "overall" mesh displacement $(d_1^T, ..., d_{n_V}^T) \in \mathbf{R}^{3n_V}$ as $d\mathbf{v}$. We refer to the set $v + S$ as *set of candidate locations* of v. Sets of candidate displacements can also be defined as vertex-individual sets: In this case we denote the set of candidate displacements for vertex v_i as $L_i \subset \mathbf{R}^3$, or, if it adds to clarity, also as $L(v_i)$.

Whenever we denote sets of candidate displacements in contexts that do not require an explicit specification as to whether we refer to global sets S or vertex-individual sets L_i, we denote sets of candidate displacements as S for ease of notation.

Discrete sets of candidate displacements are also denoted as *lists*, $S := \{s_i\}_{i=1}^{n}$, where n is the number of candidate displacements. For a discrete set of candidate displacements, we denote the minimum Euclidean distance between unequal displacements $s_i, s_j \in S$ as *sampling distance* $\delta_S := min_{s_i \neq s_j} \|s_i - s_j\|$. The sampling distance δ_{L_i} of a vertex-individual set L_i is defined analogously. In case δ_{L_i} is equal for all $v_i \in V$, we refer to it as δ_L. Discrete sets of candidate displacements yield discrete sets of candidate locations per vertex, $v + S$. In this case we refer to a candidate location $v + s \in v + S$ also as *sample point*.

A displacement field induces a deformation of all points on a mesh $M := (V, E, F)$ by means of their barycentric coordinates. This deformation can be described as a well-defined function on $\mathcal{F}$, namely $\mathrm{id}_{\mathcal{F}}$. For an intersection-free mesh M, we can denote a mesh deformation as a well-defined function m on $\mathfrak{M}$: With $\hat{v}_i := v_i + d_i$ denoting the displaced vertex positions, $\hat{V} := \{\hat{v}_i\}$, and $\hat{M} := (\hat{V}, E, F)$,

$$\mathrm{m} : \mathfrak{M} \to \hat{\mathfrak{M}}, \alpha v_j + \beta v_k + \gamma v_l \mapsto \alpha \hat{v}_j + \beta \hat{v}_k + \gamma \hat{v}_l \ .$$

2.2.2 Appearance Cost

We refer to a function $\phi : V \times S \to \mathbf{R}_0^+$ that assigns scalar *appearance costs* to pairs of a vertex and a candidate displacement as *appearance cost function*. Appearance costs reflect the dissimilarity between actual image appearance in the vicinity of candidate locations on the one hand, and expected intensities as captured by an appearance model on the other hand. For vertex-individual sets of candidate displacements L_i the respective cost function is defined as $\phi : \{(v_i, l) : v_i \in V, l \in L_i\} \to \mathbf{R}_0^+$. In case only the "current" vertex positions of a mesh are of interest, i.e. $S = \{\mathbf{0}\}$, we denote appearance cost functions as $\phi : V \to \mathbf{R}_0^+$.

A *global* appearance cost function $\Phi : \mathbf{R}^{3n_V} \to \mathbf{R}_0^+$ measures the dissimilarity between actual image appearance in the vicinity of a surface mesh depending on its shape vector $\mathbf{v} \in \mathbf{R}^{3n_V}$ and expected intensities as captured by an appearance model. In this thesis we employ global appearance cost functions which are sums of individual vertex-wise costs (cf. e.g. Sec. 3.3, and 4.2.2). Formally, this means $\Phi(\mathbf{v}) := \sum_{v \in V} \phi(v)$. This is a simple yet common definition (see e.g. Cootes et al. (1995); Khoshelham (2007); Yin et al. (2010)). While global appearance cost functions may be defined differently, e.g. by means of mutual information as is popular for multi-modal image registration (see e.g. Modersitzki (2004)), a respective discussion lies out of the scope of this thesis.

Chapter 3

Deformable Meshes for Automatic Segmentation

Contents

In this work, we build upon SSM-based segmentation, as introduced in Section 1.2.1. We describe the respective methodology in Section 3.1.

Apart from a model of the shape of an anatomical structure, another core component of the Deformable Model approach is a model of the structure's *appearance* to capture intensity patterns inside or at the boundary or around the structure as yielded by the imaging modality of interest. While promising approaches for automated learning of specific appearance models from training images exist (see e.g. (Heimann and Meinzer, 2009)), we stick to manually tailored application-specific appearance models in this thesis: Their simplicity allows for understanding (and debugging) every single step of the respective mesh deformation process, and hence facilitates the development and improvement of deformation models. Furthermore, such appearance models have been shown to yield comparatively very accurate segmentation results in practical applications (cf. e.g. Sections 7.1 and 8). We refer to such models as *heuristic* appearance models. The basics are described in Section 3.2.

SSMs as proposed by Cootes et al. (1995) are *point distribution models*: A shape is represented by a *point cloud* (see also Section 3.1.1). Triangular meshes are a straight-forward option to make such point clouds into surfaces, for which one can then compute important shape properties such as inside and outside, normals, curvature, etc. Apart from their direct link to SSMs, meshes are advantageous in many respects among a range of different types of shape representations, such as flexibility and topology preservation (Montagnat et al., 2001). Furthermore, *anatomical correspondences* (cf. Section 3.1.2) are approximately maintained during mesh deformations. Another advantage of meshes is that for particular deformation models, discrete optimization methods exist that can find *optimal* deformations w.r.t. the match between an appearance model and actual image appearance (cf. Sections 3.4.2, 3.5 and Chapter 4). We focus on deformable meshes in this thesis. For a discussion of alternative shape representations, we refer the interested reader to (Montagnat et al., 2001).

SSMs capture shape variability as contained in a training set. In general, an SSM cannot accurately represent individual shapes not used for model training. This motivates the need for deformation models that are more flexible than SSMs, i.e. capture a larger range of deformations, yet maintain specific shape knowledge (Heimann and Meinzer, 2009). To gain this kind of flexibility, in this thesis, we employ *shape constrained free deformations* of meshes after SSM-based segmentation. While the term *shape constrained free* may sound like an oxymoron at first sight, our notion of the term "free" resolves the apparent contradiction: We call deformation models for triangle meshes *free* if each vertex of the mesh has its own degree(s) of freedom, and regularization is imposed only by means of *local* criteria, e.g. criteria that apply to vertex displacements in local neighborhoods of a vertex. We refer to a free deformation model as *shape-constrained* if it imposes local constraints on displacements of vertices that somehow "maintain" the shape resulting from SSM-based segmentation. Shape constrained free deformation models are described in Section 3.4.

Free deformations can result in inaccurate segmentations in regions where the sought anatomy cannot be distinguished (by means of image appearance) from adjacent structures in the image data. In consequence, when adjacent structures are segmented separately, overlapping segmentations can occur. To avoid such inaccuracies and overlapping segmentations, *simultaneous free deformation* of multiple, *coupled* surface meshes can be performed. Section 3.5 describes a shape-constrained free multi-object deformation model which is able to simultaneously deform multiple adjacent shapes while respecting constraints on their distance. In Section 3.5.2 we contribute a *mesh-coupling method* which makes simultaneous multi-object deformation applicable to adjacent shapes of arbitrary geometry and guarantees non-overlapping segmentation results under certain conditions.

The deformation models and search algorithms described in this chapter are building blocks that can be assembled to form pipelines for fully automatic segmentation. In Chapters 7 and 8 we tailor application-specific segmentation methods from these building blocks. What the present Chapter does not provide is an exhaustive overview or description of alternative methods. In this regard we refer the interested reader to the literature in the respective Sections.

3.1   Statistical Shape Models (SSMs) for Segmentation

This section describes a framework for fully automatic segmentation of anatomical structures based on statistical shape models (SSMs). An SSM is a deformation model that captures the variability of an anatomical structure's shape among a population of individual *training shapes*. Section 3.1.1 describes the computation of such a model. Before model computation, *shape correspondences* have to be established for a training set of individual shapes. This is the subject of Section 3.1.2. Sections 3.1.3-3.1.4 describe how to employ an SSM for image segmentation.

Segmentation by deformation of an SSM exploits prior knowledge about the typical shape of the sought anatomy: Any segmentation resulting from SSM deformation is an instance of the shape model. This highly anatomy-specific deformation model renders SSM-based segmentation more robust w.r.t. imaging deficiencies as compared to conventional low-level algorithms (Heimann and Meinzer, 2009). Moreover, SSMs can deal explicitly with missing image features through their ability to *extrapolate* the respective regions of the shape from regions that are better distinguishable in terms of image features. This property can be exploited further by means of *compound SSMs* for extrapolation-based segmentation of structures that are themselves barely distinguishable, but located in the vicinity of a better distinguishable structure. The extrapolation capabilities of SSMs are discussed in Section 3.1.5.

Overall, the framework of methods described in this section allows for fully automatic segmentation of anatomical structures that can be modeled with SSMs and exhibit distinct image features in the image data (i.e. are somehow "visible"). We employ these methods throughout the Application Chapters 7-10. Alternatives for individual methods can be found in the survey on SSMs for segmentation by Heimann and Meinzer (2009).

3.1.1 Generation of SSMs

A *point distribution model* or *statistical shape model* (SSM) as proposed by Cootes et al. (1995) is computed by performing principal component analysis (PCA, see e.g. Jolliffe (2005)) on a *training set* of shape vectors, $\mathbf{T} := \{\mathbf{v}_i \in \mathbf{R}^{3n_V} | i = 1 \ldots n_{\mathbf{T}}\}$, where $n_{\mathbf{T}}$ is the number of training data. The respective vertex sets V_i each sample the surface of an individual *training shape*. Section 3.1.2 explains how to generate these training shape vectors. PCA on the training set yields a linear model

$$\overline{\mathbf{v}} + \sum_{k=1}^{m} b_k \cdot \mathbf{p}_k \tag{3.1}$$

where $\overline{\mathbf{v}}$ represents the mean shape vector, i.e. $\sum \mathbf{v}_i / n \in \mathbf{R}^{3n_V}$, $\mathbf{p}_k \in \mathbf{R}^{3n_V}$ are the *modes of shape variation*, i.e. the eigenvectors of the covariance matrix $C = \sum (\mathbf{v}_i - \overline{\mathbf{v}})(\mathbf{v}_i - \overline{\mathbf{v}})^T / n_{\mathbf{T}}$ with non-zero eigenvalue, m is the number of such modes, and $b_k \in \mathbf{R}$ are variable *shape weights*. The covariance matrix C is symmetric and its rank is at most $min(n_{\mathbf{T}} - 1, 3n_V)$. Hence it has at most $min(n_{\mathbf{T}} - 1, 3n_V)$ non-zero eigenvalues (see e.g. Fischer (2005)). In most practical applications $min(n_{\mathbf{T}} - 1, 3n_V) = n_{\mathbf{T}} - 1$, because the training set of shapes is usually much smaller than the number of point positions used to sample a shape's surface. For applications discussed in this thesis, $n_{\mathbf{T}} < 200$ and $3n_V > 20000$. Note that vectors $\mathbf{p}_k$ are not unique in case C has a non-zero eigenvalue whose algebraic multiplicity exceeds 1 (see e.g. Fischer (2005)). In this case, any set of orthogonal vectors that forms a basis of the respective eigenspace can be chosen as $\mathbf{p}_k$. With $P := (\mathbf{p}_1, ..., \mathbf{p}_{n_{\mathbf{T}}-1})$ denoting the matrix assembled from the eigenvectors, and $\mathbf{b} := (b_1, .., b_{n_{\mathbf{T}}-1})^T$ the variable *shape weight vector*, the model (3.1) becomes $\overline{\mathbf{v}} + P \cdot \mathbf{b}$. We denote the shape vector represented by a shape weight vector $\mathbf{b}$ by means of (3.1) as *shape instance* $\mathbf{v}(\mathbf{b})$, its set of vertices as $V(\mathbf{b})$, and its vertex v_j as $v_j(\mathbf{b})$. Any training vector can be represented by a particular shape weight vector: $\forall \mathbf{v}_i \in \mathbf{T} : \exists \mathbf{b} \in \mathbf{R}^{n_{\mathbf{T}}-1} : \mathbf{v}_i = \mathbf{v}(\mathbf{b})$.

A transformation $T \in \mathbf{R}^{4n_V \times 4n_V}$ is included in the model to describe transformed shapes. T is composed of an affine transformation $t \in \mathbf{R}^{4 \times 4}$ (usually rigid or rigid plus uniform scale) and zero matrices $\mathbf{0} \in \mathbf{R}^{4 \times 4}$,

$$
T = \begin{pmatrix} t & \mathbf{0} & \cdots & \mathbf{0} \\ \mathbf{0} & t & \cdots & \mathbf{0} \\ \vdots & \vdots & \ddots & \vdots \\ \mathbf{0} & \mathbf{0} & \cdots & t \end{pmatrix} .
$$

Assuming (without a change of notation) that vertex positions $v_j(\mathbf{b})$ are represented in homogeneous coordinates, the model becomes

$$
\mathbf{v}(\mathbf{b}, T) := T(\mathbf{v}(\mathbf{b})) = T(\overline{\mathbf{v}} + P \cdot \mathbf{b}) \tag{3.2}
$$

The shape weights $\mathbf{b} \in \mathbf{R}^{n_{\mathrm{T}}-1}$ and the transformation t constitute the degrees of freedom of the model.

The eigenvectors $\{\mathbf{p}_k\}_{k=1}^{n_{\mathrm{T}}-1}$ span the same vector space as the training vectors' differences to the mean, i.e. $\{\mathbf{v}_k - \overline{\mathbf{v}}\}_{k=1}^{n_{\mathrm{T}}}$. However, the eigenvectors are *sorted* by eigenvalues $\lambda_1 \geq \lambda_2 \geq \dots \geq \lambda_{n_{\mathrm{T}}-1} \geq 0$, i.e. by variance. This can be used for *dimensionality reduction* (see e.g. Heimann and Meinzer (2009)): It is possible to *approximate* every training shape by a linear combination of only the first c eigenvectors. The number c can be chosen such that the accumulated variance $\sum_{k=1}^{c} \lambda_k$ "explains" a certain percentage of the total variance $\sum_{k=1}^{n_{\mathrm{T}}-1} \lambda_k$ (see Cootes and Taylor (2004); Heimann and Meinzer (2009)). Furthermore, each b_k can be restricted to a "plausible" interval, commonly $b_k \in \left[-3\sqrt{\lambda_k}, 3\sqrt{\lambda_k}\right]$, where the idea is to assume independent distributions of each mode and constrain the respective shape weights to three standard deviations (Cootes and Taylor, 2004; Heimann and Meinzer, 2009). Alternatively, b_k can be restricted to an interval formed by the respective minimum and maximum shape weight that appears in the training set. In this thesis we stick to the latter option because it guarantees coverage of all training shapes without making any assumptions on the distribution of shape weights.

3.1.2 Prerequisites: Shape Correspondences and Alignment

Building an SSM requires a set of training surfaces. These surfaces can be generated from manual segmentations of individual tomographic image data (cf. Sec. 2.1.4). Corresponding point positions, also called *landmarks*, need to be identified on all training surfaces. Points sampled on different training surfaces need to correspond not only by means of their number (i.e. the dimensions of the shape vectors), but also in an *anatomical* sense: E.g. for training surfaces of the pelvic anatomy, if point v_k of training vector $\mathbf{v}_i$ lies on the tip of the tailbone of the i-th individual pelvis, then point v_k of training vector $\mathbf{v}_j$ has to lie on the tip of the tailbone (of the j-th individual pelvis), too.

As with segmentations of anatomical structures (cf. Sec. 1.1.3), in general, "true" anatomical correspondences are not known, and manually identified reference landmarks serve as a substitute. While the accuracy of landmark placement is subject to thorough evaluation (Heimann and Meinzer, 2009), it does not lie within the scope of this thesis.

We aim at a dense, uniform (i.e. as equidistant as possible) placement of landmarks on training surfaces. Uniform landmark placement serves for an "equal influence" of all regions of a shape on the shape modes resulting from PCA. This is desirable in case there are no specific prior assumptions on shape variance. Dense landmark placement serves for "good" approximation of original training surfaces by means of resulting point clouds and respective triangle surface meshes. As a rule of thumb, we aim at minimal landmark distances of the same order of magnitude as voxel distances of images to be segmented. Computing such dense *point-to-point* correspondences of training surfaces is considered the most challenging part of generating an SSM (Heimann and Meinzer, 2009).

In this thesis, to establish point-to-point correspondences between training surfaces, we follow the method described by Lamecker et al. (2002, 2004a): Based on significant anatomical and geometrical features, all training surfaces are consistently sub-divided into *patches*, i.e. regions on the surface with the topology of a disk. Patches are identified in a semi-automated, i.e. partly manual, process. Point to point correspondences are established for all training shapes by mapping each individual patch to a disk. As a result all training shapes can be represented in a common vector space $\mathbf{R}^{3n_V}$, with n_V the number of landmarks used to discretize the shapes, and can be equipped with corresponding triangulations to make them into surfaces. The approach of Lamecker et al. (2002, 2004a) can handle surfaces with arbitrary topology, and allows for manual definition (and hence control over the location) of some distinct landmarks, as e.g. positions of characteristic geometric features of an anatomical structure like salients or notches. Such distinct landmarks are also called *anatomical landmarks*.

After establishing point-to-point correspondences, shapes must be aligned into a common frame of reference by rigid (optionally plus scale) transformation. The most popular method for this task (Heimann and Meinzer, 2009) is the generalized Procrustes alignment (Goodall, 1991; Gower, 1975), which we also apply for generation of the SSMs employed in this thesis as described in the respective Application sections in Chapters 7-8.

3.1.3 Image Segmentation via SSM Deformation

Segmentation using an SSM (3.2) is the task of finding the set of transformation and shape parameters $(\mathbf{b}, T)$ such that $\mathbf{v}(\mathbf{b}, T)$ approximates the (unknown) shape of the sought anatomy as good as possible. An *appearance model* (cf. Sec. 3.2) encodes, based on prior knowledge, how the sought anatomy is expected to look like in the image data. A *cost function* $\Phi : \mathbf{R}^{3n_V} \to \mathbf{R}$ measures how well the *actual* gray-value appearance of a shape instance $\mathbf{v}(\mathbf{b}, T)$ in the image data matches with the *expected* gray-value appearance as captured by the appearance model. With the *objective* Φ, the task of segmenting an anatomical structure in medical image data by means of an SSM can be formulated as an optimization problem:

Find $(\mathbf{b}, T)$ for which $\Phi(\mathbf{v}(\mathbf{b}, T))$ is minimal.   (3.3)

This formulation of the segmentation problem is based on the assumption that Φ is a suitable measure of how well a surface mesh represents the sought anatomy's "true" shape (cf. Sec. 1.2.4).

In practical applications, the objective Φ is usually highly non-convex. By far the most popular optimization method is the Active Shape Model (ASM) algorithm introduced by Cootes et al. (1995). Though the field of numerical optimization provides a range of "standard" optimization techniques which have been analyzed much more thoroughly than ASM concerning theoretic properties like guaranteed convergence, ASM is generally faster and segmentation results are not known to be any less accurate (Heimann and Meinzer, 2009). We stick to ASM in this thesis and describe the approach in the following. Note in advance that ASM depends on rough initialization of the SSM's pose via a rigid (plus scale) transformation T. Automated methods for pose initialization are discussed in Section 3.1.4.

With the goal of solving (3.3), ASM as proposed by Cootes et al. (1995) perform iterative local search for appearance match as specified in Algorithm 1.

Algorithm 1 Active Shape Models (ASM)

1: $\mathbf{v}^0 := \mathbf{v}(0, T^0)$. The computation of T^0 is described in Section 3.1.4.
2: Compute an *appearance match* displacement field $d_a : V^i \to S$ that assigns a displacement vector to each vertex of the current mesh $\mathbf{v}^i := \mathbf{v}(b^i, T^i)$. The purpose of d_a is to temporarily deform the mesh onto image features. Computation of d_a is described in Section 3.3.
3: *Project* the overall mesh displacement $d_a \mathbf{v}^i$ onto the SSM. This means solving the following optimization problem: *Find* $\mathbf{b}, T$ *for which* $|(\mathbf{v}^i + d\mathbf{v}^i) - \mathbf{v}(\mathbf{b}, T)|^2$ *is minimal.* This is done subsequently for T and $\mathbf{b}$, where each sub-problem has a closed-form solution: T^{i+1} is determined for fixed $\mathbf{b}^i$ (see e.g. Eggert et al. (1997) for respective algorithms). Then, $\mathbf{b}^{i+1}$ is determined from the residual displacements remaining after application of T^{i+1} by solving an overdetermined system of linear equations (cf. Sec. 3.1.5). The resulting "effective" displacement field is $d : V^i \to \mathbf{R}^3, v \mapsto v(\mathbf{b}^{i+1}, T^{i+1}) - v(\mathbf{b}^i, T^i)$.
4: Update $i \leftarrow i + 1$ and return to step (2) unless *convergence* has been achieved. The convergence criterion is $|\mathbf{v}(\mathbf{b}^i, T^i) - \mathbf{v}(\mathbf{b}^{i+1}, T^{i+1})| > n_V \cdot \epsilon$ with a user defined threshold ϵ. In case of convergence, return $\mathbf{b}^* = \mathbf{b}^i$ and $T^* = T^i$.

A common way of profiting from an SSMs dimensionality reduction capability is to perform multiple runs of ASM in a *hierarchical, multi-level* manner, starting with a first run with just a few shape weights $\{b_1, ..., b_{n_1}\}$, $n_1 < n_T - 1$, and iteratively conducting further runs with more and more shape weights $n_2 < n_3 < ... \leq n_T - 1$, each initialized with the result of the previous run.

Note that in the process of mesh deformation via ASM, for each vertex v, the resulting vertex displacement $d(v)$ as computed in Step 3 may deviate arbitrarily far from the respective candidate displacement chosen for appearance match, $d_a(v)$, as computed in Step 2. In general, for mesh deformations by means of *global* deformation models like rigid transformations or SSMs, sets of candidate displacements (which are assessed for appearance match, cf. Sec. 3.2) and sets of potential resulting displacements, i.e. the *ranges of motion* of vertices, are not closely related. This is different for *free* mesh deformations, as discussed in Section 3.4 and Chapter 4.

3.1.4 Initial Shape Detection

A well-established, reliable and widely used automatic approach for global detection of large, well-distinguishable anatomical structures with limited shape variability (e.g. bones) is the *Generalized Hough Transform* (GHT, see Ballard (1981); Khoshelham (2007)): Given a shape vector $\overline{v} \in \mathbf{R}^{3n_V}$ representing the average shape of an anatomy, together with some position and orientation (i.e. a rigid transformation T), an appearance cost is determined by comparing image gradient directions $\frac{\nabla I(v)}{|\nabla I(v)|}$ and surface normals n_v at each vertex v of the transformed surface mesh. If the angle between image gradient and vertex normal is smaller than a threshold, the individual vertex is assigned zero cost, else one. I.e.,

$$\phi_{GHT} : V \to \{0, 1\}, v \mapsto \left(\arccos(\frac{\nabla I(v) \cdot n_v}{|\nabla I(v)|}) < \frac{\pi}{a} \right) ,$$

where $\frac{\pi}{a}$ defines an angle threshold. The sum of vertex costs determines the (overall) image cost $\Phi_{GHT}(T(\overline{v}))$.

Exhaustive search for a globally optimal T is possible at least over some coarse discretization of the degrees of freedom of T: The surface mesh is put at every position and orientation (at every scale) in the image. The respective image cost is approximated in a computationally efficient manner by sorting surface normals and gradient directions into "bins" of angle pairs in a spherical coordinate system, and counting vertices which are sorted into the same bin as the respective image gradient. Bins span angles $\frac{\pi}{a}$ in both angle coordinates. The (or one of the) T with minimal image cost serves as resulting initial pose of an SSM. In practice, $a = 4$ has proven to provide for *fail-safe* initial shape detection: It yields initial transformations that put sought structures "within reach" of *image probes* that guide subsequent SSM deformation (cf. e.g. Sec. 7.2). Note that Φ_{GHT} is not continuous, yet this is not a practical issue due to the particular optimization approach that performs exhaustive search over a finite set of transformations.

See e.g. (Ruppertshofen et al., 2011) for references to alternative methods for pose initialization.

3.1.5   Lack of Image Features: SSMs for Extrapolation

Step 3 of the ASM algorithm (Sec. 3.1.3) computes SSM parameters $(\mathbf{b}, T)$ which approximate a mesh displacement $d_a \mathbf{v}$ that moves each vertex to a position in the image which matches an appearance model. Parameters $\mathbf{b}$ and T are determined subsequently. In the following, we have a closer look at how $\mathbf{b}$ is computed. We are looking for a shape weight vector $\mathbf{b}$ that describes the displaced mesh: $\mathbf{v}(\mathbf{b}, T^i) = (\mathbf{v}^i + d_a \mathbf{v}^i)$. With $A := T^i \cdot P$ and $\mathbf{c} := (\mathbf{v}^i + d_a \mathbf{v}^i - T^i \cdot \bar{\mathbf{v}})$ this becomes $A\mathbf{b} = \mathbf{c}$. This is a system of linear equations with less than $n_\mathbf{T}$ unknowns and $3n_V$ rows. For the applications discussed in this thesis, $3n_V > 20000$ and $n_\mathbf{T} < 200$, i.e. $3n_V >> n_\mathbf{T}$, which makes the above system overdetermined. Hence most probably an exact solution $\mathbf{b}$ does not exist. However, instead, we can determine the weight vector with smallest *distance* to the displaced mesh. When considering Euclidean distance, this means we are looking for $\mathbf{b}$ for which $(A\mathbf{b} - \mathbf{c})^T (A\mathbf{b} - \mathbf{c})$ is minimal. This is a *linear least squares problem* (see e.g. Björck (1996)). Such a $\mathbf{b}$ exists and is unique if A has full rank, and is determined by solving $A^T A\mathbf{b} = A^T \mathbf{c}$.

This way, all the individual vertex distances $|v_j(\mathbf{b}, T^i) - (v_j^i + d_a(v_j^i))|^2$ have the same "influence" on the solution. In case we "are not sure" about a particular displacement $d_a(v)$, we can weigh its influence lower by multiplying it with some weight $w(v) \geq 0$. We can set a weight for each vertex, yielding a diagonal matrix $W := diag(w(v_1), ..., w(v_{n_V}))$. The optimization problem becomes to find $\mathbf{b}$ for which $(A\mathbf{b} - \mathbf{c})^T W(A\mathbf{b} - \mathbf{c})$ is minimal. This is referred to as *weighted linear least squares problem* (see e.g. Björck (1996)). Such a $\mathbf{b}$ exists and is unique if WA has full rank. It is determined by solving $A^T WA\mathbf{b} = A^T W\mathbf{c}$.

When setting $w(v) := 0$, this means vertex v is *ignored* when determining $\mathbf{b}$. In effect, its updated position $v(\mathbf{b}, T^i)$ is *extrapolated* from all other vertices that are equipped with $w > 0$. As $3n_V >> n_\mathbf{T}$, in practice, ignoring "many" points still leaves us with a WA that has full rank and hence with a unique solution $\mathbf{b}$.

This extrapolation capability of SSMs can be exploited in various ways. First, appearance match can be analyzed not only to assign costs to candidate displacements per vertex (cf. Sec. 3.3), but also to assign a *weight* to each vertex. If no candidate displacement yields a match with an appearance model, the weight can be set to zero. Second, imagine a structure that exhibits only weak or no image features, yet its shape correlates with the shape of a better visible structure. In this case, a *compound SSM* can be generated that contains both structures. Then, the better visible structure is deformed onto the image data by means of appearance match, while the low-contrast structure is purely extrapolated. This requires the well-visible structure to have more than n_T vertices. An exemplary application for this usage of compound SSMs is presented in Section 7.3.

3.2 A Simple Heuristic Appearance Model

An appearance model captures prior knowledge about the gray value characteristics that the sought anatomy exhibits when imaged with a particular imaging modality. Hence in contrast to deformation models, appearance models are imaging modality specific.

A very simple but powerful heuristic appearance model assumes that each point $x \in \mathbf{R}^3$ on the target structure's surface exhibits an image intensity $I(x)$ within a certain window $[t_1, t_2]$. Furthermore, the directional image derivative $\nabla_{n_x} I(x)$ along the target structure's surface normal n_x is expected to exceed a certain threshold $g > 0$, or under-run a threshold $-g < 0$, or either of the two, depending on the application. Appearance costs can be derived from this appearance model as described in Sec. 3.2.1. If necessary for a particular application, the parameters t_1, t_2, g can be estimated automatically from individual image data, as described in Sec. 3.2.2.

The appearance model described in this section is the basis for most applications described in Chapters 7-10. Application-specific modifications of the above are described in the respective Application sections.

3.2.1 Appearance Cost Function

For a range of applications discussed in this thesis, we define vertex-individual appearance costs by means of an *appearance cost function* $\phi : V \times S \to \mathbf{R}_0^+$. The function ϕ reflects the "match" between appearance model and image $I : \Omega \to \mathbf{R}$ at vertex candidate locations $v + s$ as follows: If the image intensity $I(v + s)$ lies within a window $[t_1, t_2]$, and the directional image derivative $\nabla_{n_v} I(v + s)$ along the surface normal n_v at vertex v exceeds a threshold $g > 0$, the cost $\phi(v, s)$ is proportional to $\frac{g}{\|\nabla_{n_v} I(v+s)\|}$. Otherwise costs are set to a constant, high value ϕ_{high}. Depending on the application (and denoted in the respective Application sections), what is expected for the directional derivative can also be $\nabla_{n_v} I(v + s) < -g < 0$ or $\|\nabla_{n_v} I(v + s)\| > g > 0$. The thresholds t_1, t_2 and g are parameters of the strategy and are set per application (as described in the respective application sections) either to fixed values or to image-individual values that are determined via histogram analysis (cf. Sec. 3.2.2).

A simpler alternative is a binary function $\phi : V \times S \to \{0, 1\}$ that decides whether a vertex candidate location matches or not. Such a function ϕ can also be designed such that it chooses no more than one displacement to match within the respective set of candidate displacements, i.e. $\forall v \in V : \exists_1 s \in S : \phi(v, s) = 1$. An exemplary application is presented in Section 7.1.

3.2.2 Intensity Parameter Estimation

For some applications, there is no need to estimate image-individual parameters t_1, t_2, g, as e.g. applications on CT data which is *calibrated*, i.e. corresponding tissues of different individuals exhibit similar intensity characteristics.

Other applications (e.g. involving contrast-enhanced CT or MRI) call for image-individual parameters. If the intensity distribution inside the sought anatomy can be modeled as a Gaussian mixture, and $[t_1, t_2]$ encloses a distinguished "peak" of the respective Gaussians (e.g. the darkest, brightest, highest...), the intensity window can be estimated by fitting a Gaussian mixture model to a histogram of image intensities via the Expectation Maximization (EM) algorithm (McLachlan and Krishnan, 2007). The histogram can e.g. be computed from image voxels in and/or around an initial mesh. Parameter estimation can be repeated after some intermediate steps of model deformation, with the goal of getting more precise estimates. As for the gradient threshold g, image-individual values can be estimated based on the assumption that tissue appears at least to some extent homogeneous in the image data. Then, g can be set to some threshold above which a low percentage of gradient magnitudes appear in a histogram. A histogram of gradient magnitudes can be assessed over the whole image or over some band around the initial template shape.

3.3 Local Search for Appearance Match

Section 3.3.1 discusses the geometry of sets of candidate displacements. We describe a specific geometry which is commonly employed in the context of all methods for mesh deformation introduced in this chapter. An alternative geometry will be discussed in Chapter 4. Section 3.3.2 discusses how to derive an appearance match displacement field on the basis of sets of candidate displacements and an appearance match cost function. In the context of the appearance cost function described in Section 3.2.1 and sets of candidate displacements as specified in Section 3.3.1, we employ the term *intensity profile*, which is introduced in Section 3.3.3.

3.3.1 Unidirectional Displacements

Unidirectional, i.e. linear, one-dimensional sets of candidate locations per vertex of a deformable mesh are commonly employed (Heimann and Delingette, 2011) due to a number of benefits: (1) Appearance match assessment is fast; (2) It is easy to select the "best" candidate location: A one-dimensional set of candidate locations is likely to hit the target surface at only one single point (or at most a finite number of points), and hence the set of candidate locations with matching appearance is likely to be small. Further benefits of unidirectional sets of candidate locations that particularly apply in combination with free deformation models will be discussed in Section 3.4. Downsides of unidirectional sets of candidate locations are discussed in Chapter 4.

Unidirectional sets of candidate locations lie along *displacement directions* per vertex v_i of a surface mesh, which we denote as $\ell_i \in \mathbf{R}^3$, where $\|\ell_i\| = 1$. If it adds to clarity of notation, we denote $\ell(v_i) := \ell_i$. Directions $\ell(v)$ are commonly defined to run along vertex normals n_v (Heimann and Delingette, 2011). Different directions can be chosen as well (cf. Sec. 3.5 and 9.2.1). A number of equidistant points $v_i + l$ build the set of candidate locations for v_i. The *displacement* l is an element of a vertex-individual discrete list of *unidirectional candidate displacements*

$$L_i := \left\{ l_k := (k - \frac{n_L + 1}{2}) \cdot \delta_L \cdot \ell_i \right\}_{k=1}^{n_L}$$

where δ_L is the *sampling distance* (cf. Sec. 2.2), $0.5 \cdot n_L \cdot \delta_L$ is the *reach* (or "radius"), denoted as r_L, and $2r_L$ is the *length* of the set of candidate displacements. If it adds to clarity of notation, we denote $L(v_i) := L_i$, and also $n := n_L$ and $r := r_L$. Sampling distance and probe length are parameters of the method. We set them to application-specific, heuristically determined values as specified in the respective Application sections (Chapters 7-9). The resulting set of candidate locations for a vertex v is $v + L(v)$. Note that for vertex position v to be among the set of candidate locations, i.e. for the possibility of *not* displacing a vertex, n_L has to be an odd number.

3.3.2 Optimal Displacement Fields

For each vertex v and candidate displacement $l \in L(v)$, an *appearance cost* is determined from image data by evaluating a cost function $\phi(v, l)$ (cf. Sec. 3.2.1). The minimum cost candidate displacement may serve as a desired (locally optimal) displacement for the respective vertex. This yields an appearance match displacement field

$$d_{localOpt} : V \to \bigcup_{v \in V} L(v), v \mapsto \underset{l \in L(v)}{\operatorname{argmini}} \{\phi(v, l)\} \ .$$

This displacement field minimizes $\sum_{v \in V} \phi(v, d(v))$. The appearance match displacement field $d_{localOpt}$ is the common choice for guiding SSM deformation (cf. Sec. 3.1.3), although it makes Active Shape Models (ASM) as described in Algorithm 1 sensitive to outliers (Heimann and Meinzer, 2009). For more robustness against outliers, appearance match displacement fields can also be computed by means of the methods described in Sections 3.4, 3.5 and Chapter 4. For the applications discussed in this thesis, however, we did not observe failure of ASM due

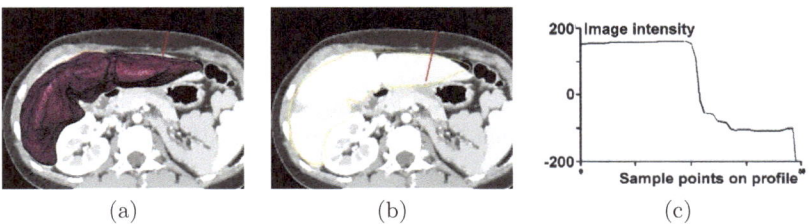

(a) (b) (c)

Figure 3.1 (a,b) Liver surface mesh (purple) with candidate locations for an exemplary vertex in slice of liver CT. Convex hull of set of candidate locations shown in red. (c) Intensity profile with domain as shown in (a,b).

to outliers that come with $d_{localOpt}$. Note, however, that for some applications, we reduce the impact of outliers in a different way, namely by means of *image filtering* (cf. Sections 7.1 and 7.3).

3.3.3 Intensity Profiles

For certain appearance cost functions, together with unidirectional sets of candidate displacement, assessment of appearance match for all candidate locations $v_i + L_i$ of a vertex v_i requires image information only at these candidate locations. In other words, appearance match for all $v_i + l \in v_i + L_i$ is derived from a sub-image at candidate locations,

$$I_{|v_i+L_i} : v_i + L_i \to \mathbf{R}, v_i + l \mapsto I(v_i + l) .$$

Such sub-images are called *intensity profiles*. In this particular but common case, with slight abuse of notation, we use the term *intensity profile* or just *profile* to also refer to the *domain* of $I_{|v_i+L_i}$, i.e. to the set of candidate locations $v_i + L_i$. For the appearance cost function described in Section 3.2.1, this is the case if displacement directions run along vertex normals, and image derivatives in vertex normal directions are approximated via finite differences. Figure 3.1 shows candidate locations for a particular vertex in surface normal direction and the respective intensity profile. See Chapter 7 for applications.

3.4 Shape-constrained Free Mesh Deformations

As described in Section 3.1.3, segmentation with an SSM reduces the "deformation search space" to the shape model, which is appropriate for producing a robust segmentation result. However, in general, new, unknown shapes are not contained in the model and therefore cannot be reached by SSM deformation. More flexible deformation models can overcome this limitation. In this section we describe surface mesh deformation models that can be applied for small, "final" deformations starting from an initial mesh resulting from SSM deformation. These models regularize mesh adaptation solely by means of local constraints. That means each vertex position of the mesh can move "freely", subject only to regularity constraints that consider its direct neighborhood. However, such deformation models must take care that shape knowledge captured by an initial mesh is not "lost", but somehow "remembered" during mesh deformation – or else the power of specific deformation models gets lost and problems like leakage appear again (cf. Sec. 1.2.1). We call such models *shape constrained free deformation models*.

The method described in Section 3.4.1 puts a narrow band around the initial surface mesh that cannot be left during free deformation. This approach has been published in (Kainmueller et al., 2007). Within this band the mesh deforms freely, with regularized mesh curvature. This approach is suitable for applications that call for relatively large free deformations after SSM deformation – an example is liver segmentation in contrast-enhanced CT (cf. Sec. 7.1).

The method described in Section 3.4.2 constrains the difference between the lengths of neighboring displacements. This approach is suited for relatively small deformations in situations where image features are ambiguous and hence outlier displacements need to be prevented. One exemplary application is segmentation of the pelvic bones in CT (cf. Sec. 7.2).

Note that another shape-constrained free deformation model as proposed by Pekar et al. (2001) constrains the differences between the edge lengths of the deformed mesh and the lengths of the respective edges of the initial mesh as resulting from SSM deformation. This interesting approach has been reported to be suitable for segmentation of complex bony structures, namely vertebrae. However, we cannot discuss it in this thesis due to time limitations.

Ideally, we are looking for the *combination* of SSM parameters $\mathbf{b}, T$ and additional mesh displacement $d\mathbf{v} \in \mathbf{R}^{3n_v}$ (i.e. a free deformation) for which

$$\Phi(\mathbf{v}(\mathbf{b}, T) + d\mathbf{v}) + \Psi(d\mathbf{v})$$

is minimal, where Ψ is some regularization cost function (compare (3.3)). In this thesis, we follow a coordinate search approach, i.e. we split this optimization problem into two smaller ones by first finding some $\mathbf{b}, T$ (with ASM, cf. Sec. 3.1.3) and successively looking for $d\mathbf{v}$ (with fixed $\mathbf{b}, T$). Alternative approaches directly relax the SSM deformation model – see Heimann and Meinzer (2009) for references. We follow the coordinate search approach for a practical reason: It gives us the freedom to alter a surface mesh (i.e. *re-sample / re-mesh* the surface) after SSM deformation. This way we can directly control initial mesh consistency and regularity, which is helpful for controlling free deformation models. For example, we can get rid of any mesh inconsistencies (like self-intersections) that might be induced by

SSM deformation and are problematic for free deformations due to flipped normals. Furthermore, there is an approach for multi-object segmentation (as described in Sec. 3.5) that *requires* to alter meshes. This is not to say that the above mentioned alternative approach cannot tackle the respective issues.

3.4.1 Free Deformation within a Narrow Band

The approach described in this section follows the basic idea of mesh regularization via a "force" that simply pulls each vertex towards the centroid of its neighbors, as first presented by Miller et al. (1991). In effect, mesh curvature is constrained. Additionally, we constrain mesh deformation to a narrow band around the initial surface to prevent too large deviations, and prevent self-intersections of the mesh. This approach has been published in Kainmueller et al. (2007).

To deform a mesh, the minimum cost displacement for each vertex is determined (cf. Sec. 3.3), truncated to some maximum length (i.e. a *stepsize*), and applied. Subsequently the mesh is regularized locally via a small displacement of each vertex toward the centroid of the respective adjacent vertices. Any displacement is confined to a narrow band, and trimmed such that it does not induce self-intersections. Algorithmically, mesh deformation is performed iteratively as specified in Algorithm 2. In the following, we refer to this approach as *FreeBand*.

Weighting of appearance- and smoothing displacement of a vertex depends on the angle they enclose (Step 3 of Algorithm 2): If appearance match pulls in the same direction as smoothing, we consider smoothing obsolete; in this case, $w_r(v) = 0$. Instead, if the two displacements pull in opposite directions, we set $w_r(v)$ to a maximum value w_r^{max}, which is a parameter of the method. From experiment, we set this parameter to 0.25. The special case that an appearance match displacement is zero while the respective smoothing displacement is not can arise for two reasons (cf. Sec. 3.3 and 3.2.1): (1) The vertex position itself yields the best local appearance match, which calls for $w_r(v) = 0$, or (2) no appearance match was found at all, which calls for $w_r(v) = w_r^{max}$. As a simple compromise, we set $w_r(v) = 0.5w_r^{max}$.

As a consequence of Step 5 of the Algorithm, if $\mathbf{v}^0$ has no self-intersections, then $\mathbf{v}^{i+1}$ also has this property.

Note that in general, FreeBand does not converge: Oscillations may follow from our simple definition of w_r (Step 3). Furthermore, our simple approach for displacement trimming to 0.9 times the distance to an intersection point yields non-convergence, too (Steps 4 and 5). This way of trimming is intended to cope with inaccuracies stemming from floating point arithmetic. If trimming "precisely" to the intersection point, such inaccuracies could compromise the guarantees that FreeBand stays within a narrow-band and does not produce self-intersection of the mesh.

Despite its theoretical deficiencies, FreeBand produces very accurate segmenta-

---

Algorithm 2 FreeBand

1: Set $\mathbf{v}^0 := \mathbf{v}^*$ as resulting from SSM deformation (ASM, cf. Sec. 3.1.3).

2: Compute two vector fields on the current surface mesh with vertices V^i:

  (1) An *appearance match* displacement field $d_a : V^i \to S$, just as in Sec. 3.1.3,

  (2) A *smoothing* displacement field $d_r : V^i \to \mathbf{R}^3$, pulling each vertex towards the barycenter of its 1-ring neighborhood.

3: Compute a *resulting* vector field as a convex combination of d_a and d_r, i.e. $d : V^i \to \mathbf{R}^3, v \mapsto w_a(v)d_a(v) + w_r(v)d_r(v)$, with $w_a(v) + w_r(v) = 1$. The vertex-individual weight $w_r(v)$ is computed as follows:

  (1) $|d_r(v)| > 0 \wedge |d_a(v)| > 0 \Rightarrow w_r(v) := 0.125 \cdot (1 - \frac{d_a(v) \cdot d_r(v)}{|d_a(v)| \cdot |d_r(v)|})$.

  (2) $|d_r(v)| > 0 \wedge |d_a(v)| = 0 \Rightarrow w_r(v) := 0.125$.

  (3) $|d_r(v)| = 0 \Rightarrow w_r(v) := 0$.

4: If a resulting vector $d(v)$ points out of the narrow band trim it such that it stays inside: Therefore, compute the intersection point x of $d(v)$ with the narrow band boundary, and re-set $d(v) := 0.9(x - v)$.

5: Set $\mathbf{v}^{i+1} := \mathbf{v}^i + d\mathbf{v}^i$. Perform this addition iteratively for all vertices, such that updated position $v^{i+1} := v^i + d(v^i)$ of an individual vertex does not produce self-intersections in the surface; else trim the respective displacement $d(v^i)$ accordingly like in Step 4.

6: Update $i \leftarrow i + 1$ and return to step (2) if stopping criterion has not been met (same ϵ as in Sec. 3.1.3). Otherwise, or if a maximum number of iterations has been reached, return $\mathbf{v}^* := \mathbf{v}^i$.

---

tions in practice (cf. Sec. 7.1). Transforming FreeBand into a numerically sound approach which has well-understood theoretical properties and which at the same time yields equally accurate segmentation results is subject to future work.

3.4.2 Free Deformation with Bounded Displacement Differences

Solitary outlier displacements can be remedied by simple means like curvature restriction as performed by the FreeBand approach. However, image features that are "misleading" for a *set* of neighboring vertices can be induced by modality-specific appearance characteristics of particular anatomical structures. See Figure 3.2a for an example. The approach described in this section follows the idea to stick to the initial shape during free deformation by putting a bound on the difference of lengths of neighboring displacements. This constraint does not only prevent noise-induced solitary outlier displacements – it can also prevent deformations onto systematically

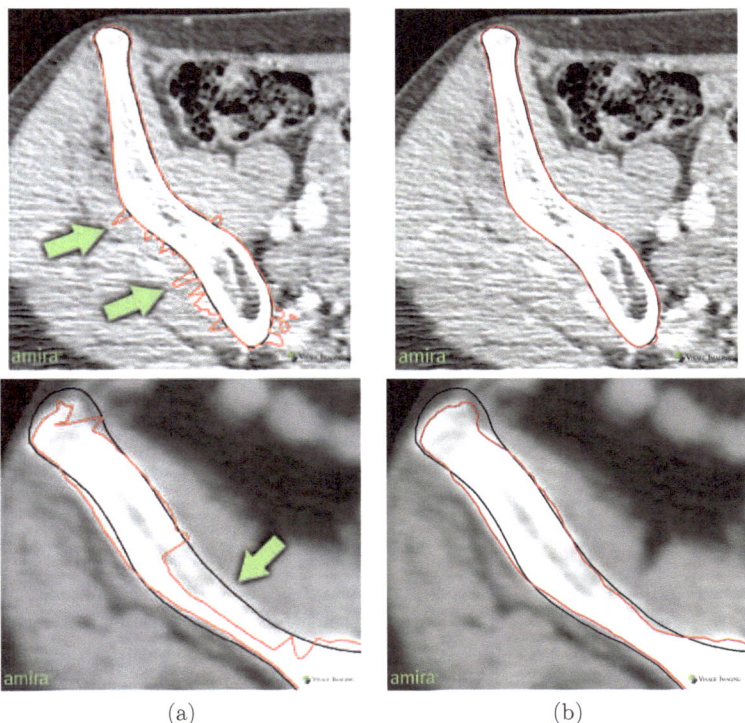

(a)   (b)

Figure 3.2 Misleading image features in exemplary slices of pelvic CT. Black: Initial meshes. Red: Deformed meshes. (a) Mesh deformation by minimum cost displacement per vertex. Top: Noise-induced outlier displacements. Bottom: Systematically mislead deformation due to ambiguous image features. (b) Deformation via *GraphCuts* as described in Section 3.4.2 results in accurate segmentation.

misleading image features, as shown in Figure 3.2b.

Formally, the optimization problem is to find a displacement field for which the sum of appearance costs is minimal while hard constraints are satisfied by the differences of lengths of neighboring displacements:

$$\text{Find } \{d_i \in L_i\}_{i=1}^{n_V} \text{ for which } \sum_{i=1}^{n_V} \phi(v_i, d_i) \text{ is minimal}$$

$$\text{subject to } \forall (j,k) \in E : |\ell_j \cdot d_j - \ell_k \cdot d_k| \leq c \; , \tag{3.4}$$

where L_i are sets of candidate displacements per vertex $v_i \in V$ in directions ℓ_i (cf. Sec. 3.3). The *shape preservation constraint* (called *smoothness constraint* by Li et al. (2006)) limits the set of feasible displacement fields to those for which lengths of displacements of adjacent vertices differ at most by some value $c \in \mathbf{R}_0^+$, where c is a multiple of the sampling delta δ_L. The smaller the parameter c is set, the more "alike" are the shapes of the initial surface and the surface resulting from optimization. Note that for this deformation model, the range of motion of a vertex *equals* its set of candidate locations. This is different for ASM and FreeBand (cf. Sec. 3.1.3 and 3.4.1)

A list of zero-displacements is a feasible solution to (3.4) for any $c \in \mathbf{R}_0^+$. As the set of feasible solutions is finite, existence of an optimal solution is guaranteed. However, an optimal solution of (3.4) is not necessarily unique. Methods as described in the following are able to find *all* optimal solutions of (3.4). Each optimal solution corresponds to a segmentation. This raises questions not only w.r.t the quality of the model and the task of evaluating segmentation results, but also concerning the practical value of "having a choice" among a set of optimal segmentations. However, as discussed in Section 1.2.4, this thesis does not pursue this path of investigations. Instead, we simply select the first optimal solution as put out by a respective algorithm.

For contours in 2D, the optimization problem (3.4) can be solved with Dijkstra's or Viterbi's optimal path searching algorithms (Dijkstra, 1959; Viterbi, 1967), which is popular for interactive or automatic 2D (slice-wise) segmentation techniques (Stalling and Hege, 1996; Mortensen and Barrett, 1998; Falcão et al., 1998; Amini et al., 1990; Schenk et al., 2000; Behiels et al., 2002). As for surfaces in 3D, more recently, Li et al. (2006) showed how a graph can be constructed for which optimal graph searching by means of min-cut/max-flow algorithms yields a global minimizer of (3.4) with computational complexity of low polynomial order. This optimization method requires the (signed) length of displacements to define a total order on the sets of candidate displacements per vertex, i.e. sets of candidate displacements must be *unidirectional* (cf. Sec. 3.3). For details on graph construction we refer the reader to (Li et al., 2006).

In the following, we refer to this method, i.e. the combination of problem formulation (3.4) and solver for surface meshes in 3D, as *GraphCuts*. GraphCuts have proven powerful for accurate fine-grain segmentation of 3D medical image data (Yin et al., 2010; Lee et al., 2010; Zhang et al., 2010; Petersen et al., 2011). Furthermore, an extension of GraphCuts allows for simultaneous segmentation of multiple objects (Yin et al., 2010), a property that we call *multi-object ability*. The latter is described in Section 3.5.

As opposed to ASM and FreeBand, GraphCuts as described in this section does not perform search for appearance match and deformation iteratively, but computes one single deformation which is optimal within a local search radius in the sense of (3.4). Hence it is suitable in situations where the sought anatomy can be assumed

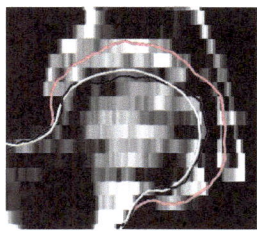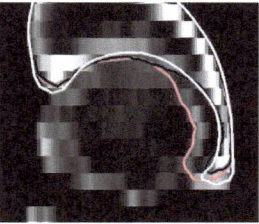

Figure 3.3 Coronal slices of pelvic CT. Details showing hip joint. White: Gold standard segmentations of femoral head (left) and acetabulum (right). Red: Segmentations obtained by single-object shape constrained free deformations with GraphCuts (cf. Sec. 3.4.2). Ambiguous image features cause inaccurate segmentations.

to be within reach along surface mesh normals (or other pre-defined directions) – i.e. for relatively small deformations.

GraphCuts can also be used in combination with ASM and FreeBand (cf. Sec. 3.1.3 and 3.4.1) to make these approaches more robust w.r.t. outlier displacements and mis-adaptations: The respective modification to ASM and FreeBand is to simply compute appearance match displacement fields d_a with GraphCuts instead of individually displacing each vertex to its minimum cost displacement. However, we do not exploit such modified ASM or FreeBand in this thesis, as discussed in Sections 3.3 and 7.1.5, respectively.

3.5   Simultaneous Free Deformations of Multiple Meshes

If adjacent structures are barely distinguishable in image data, shape-constrained free deformations of single structures can result in inaccurate segmentations. Figure 3.3 shows exemplary situations. A side-effect of inaccurate deformations is that resulting segmentations of adjacent structures can overlap if performed independently for each structure. A basic idea for refining SSM-based segmentation results and simultaneously solving the overlap problem is to deform multiple adjacent surface meshes at the same time and incorporate some knowledge about their spatial relationship.

While a range of multi-object segmentation approaches have been proposed that prevent overlap via *alternating* deformation of adjacent structures or resolve occurring overlap via collision detection (e.g. Costa et al. (2007); Gilles et al. (2006)), an extension of the GraphCuts approach (cf. Sec. 3.4.2) allows for *simultaneous* deformation of multiple adjacent objects, respecting constraints on their distance. This approach is described in Section 3.5.1. This method requires a *correspondence relation* between vertices of adjacent surface meshes, implying *shared displacement directions* that can be employed for simultaneous deformation. In Section 3.5.2 we propose a *coupling* scheme for establishing such a correspondence relation in adjacent regions of two arbitrary surfaces. We show that we can guarantee non-overlapping segmentation results under certain conditions, even in case of overlapping initial surfaces. We show resulting correspondence relations for exemplary adjacent anatomical structures.

Methods involving shared displacement directions have been described for surfaces on which corresponding vertices are easily found. This holds for height field or cylindrical surfaces in regular grids (Li et al., 2006), or if one surface can be obtained by displacing the other along its vertex normals (Li et al., 2005). Furthermore, similar to the method we propose in Section 3.5.2, Yin et al. (2008) couple adjacent surfaces with shared displacement directions. However, they do not deal with the problem that overlapping segmentation results might occur if the shared displacement directions do not fulfill certain conditions.

Liu et al. (2009) have proposed an approach which is similar to ours in that it makes use of medial axes of an initial segmentation to define displacement directions with an appropriate length while avoiding self-intersections. However, the displacement directions are specifically designed for tree-like structures with multiple walls, like airway trees and lung vascular trees. They cannot be directly used for arbitrary adjacent surface meshes.

To the best of our knowledge, our mesh-coupling method as presented in Section 3.5.2 originally contributes a method that generates shared displacement directions for arbitrary adjacent surfaces that are guaranteed to prevent overlapping segmentation results under certain conditions.

3.5.1 Multi-object Graph-based Deformation of Coupled Meshes

Given two surface meshes, $M_1 := (V_1, E_1, F_1)$ and $M_2 := (V_2, E_2, F_2)$, we are looking for a compound, multi-object displacement field

$$d : V_1 \cup V_2 \to \mathbf{R}^3, v \mapsto d(v) \in L(v) .$$

We assume that meshes are *coupled* by means of a bijective mapping of subsets of vertex indices. With respective sub-lists of vertices denoted as $U_1 \subset V_1$ and $U_2 \subset V_2$, we sloppily denote the index mapping as m : $U_1 \to U_2$. Section 3.5.2 explains how to establish such a mapping. The mapping m defines displacement directions: $\forall v \in U_1 : \ell(v) := \text{sign}(v)\frac{\text{m}(v)-v}{|\text{m}(v)-v|}$, and vice-versa for all $w \in U_2$ (cf. Sec. 3.3 for notation), where $\text{sign}(v) := 1$ if vertex v lies *outside* the surface $\mathfrak{M}_2$, and $\text{sign}(v) := -1$ otherwise, i.e. in case of initial overlap of the surfaces at vertex v. Assuming mapped vertices v and $\text{m}(v)$ to be either both inside or both outside the respective other surface, displacement directions satisfy $\ell(v) = -\ell(\text{m}(v))$. In other words, mapped vertices *share* their displacement directions. Unmapped vertices, i.e.

vertices in $V_1 \backslash U_1$ and $V_2 \backslash U_2$, are equipped with individual displacement directions, e.g. $\ell(v) := n_v$. For such coupled meshes with shared displacement directions, mesh deformation can be formulated as the following optimization problem:

$$\text{Find } \{d_i \in L_i\}_{i=1}^{n_{V_1 \cup V_2}} \text{ for which } \sum_{i=1}^{n_{V_1 \cup V_2}} \phi(v_i, d_i) \text{ is minimal and}$$

$$\forall (j,k) \in E_1 \cup E_2 : |\ell_j \cdot d_j - \ell_k \cdot d_k| \leq c, \tag{3.5}$$

$$\forall v \in U_1 : c_0 \leq |m(v) - v| - \ell(v) \cdot (d(v) - d(m(v))) \leq c_1$$

This extends the optimization problem (3.4) by distance constraints (last row) that demand distances between displaced vertex positions of mapped vertices to be within a range $[c_0, c_1]$, with $c_0, c_1 \in \mathbf{R}$. Note that in general, as opposed to (3.4), the additional constraints compromise existence of a solution to (3.5). In case of existence, a globally optimal solution can be computed via construction of a graph and application of minimum-cut/maximum-flow algorithms, as detailed by Li et al. (2006). This algorithm also detects if a feasible solution does not exist. However, we did not encounter this case in practical applications (as presented in Chapter 8).

In the following, we refer to this method as *multi-object GraphCuts*.

3.5.2 Coupling Adjacent Surface Meshes

In this section we propose a construction scheme for shared displacement directions for two arbitrary surface meshes. Shared displacement directions yield line segments along which vertices of both surfaces are displaced during mesh deformation. Our construction scheme establishes a bijective mapping between regions on the surfaces. We refer to this construction scheme as *mesh coupling* method. It has been published in (Kainmueller et al., 2009d). The mesh coupling method renders multi-object GraphCuts as described in Section 3.5.1 applicable to arbitrary adjacent surface meshes, and deformed meshes are guaranteed not to overlap in coupled regions.

In the following, we discuss necessary and sufficient conditions for overlap-free deformations of adjacent surface meshes along shared displacement directions. Subsequently, we propose a construction algorithm for shared displacement directions on adjacent surface meshes.

Conditions for Non-overlapping Deformations

As before in Section 3.5.1, we denote shared displacement directions by means of a bijective mapping of lists of vertices, $m : U_1 \rightarrow U_2$. The respective displacement direction for a vertex $v \in U_1$ is defined as $\ell(v) := \text{sign}(v) \frac{m(v) - v}{|m(v) - v|}$

We do not allow surfaces to completely "swap sides" along shared profiles. Consequently, a necessary condition for deformations of coupled surface meshes to avoid

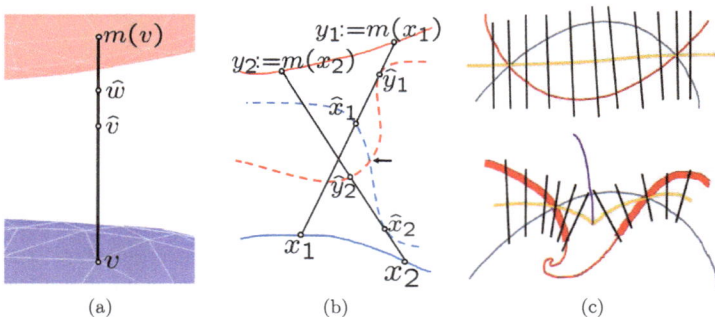

(a)   (b)   (c)

Figure 3.4 (a) Necessary condition for non-overlapping deformations: $\hat{v}$ lies closer to v than $\hat{w}$. (b) Exemplary overlapping deformations in 2D. Continuous red and blue lines: Initial contours. Black lines: Connections of mapped points intersect. Dotted red and blue lines: Exemplary deformed contours, intersecting at black arrow. (c) Exemplary situations in 2D. Red, blue: contours to be coupled. Yellow: mid contour. Purple: Skeleton of mid contour. Black: shared displacement directions. Top: Overlapping regions can be entirely mapped. Bottom: Overlapping regions cannot be entirely mapped. The thin part of the red contour remains unmapped

overlap is the following:

> For all $v \in U_1$ with corresponding $w := \mathrm{m}(v) \in U_2$,
> the displaced position $\hat{v}$ of v must lie closer to v in profile direction
> than the displaced position $\hat{w}$ of w, i.e.
> $$(v^* - v) \cdot \ell(v) < (w^* - v) \cdot \ell(v) \ .$$
> (3.6)

This condition corresponds to $c_0 > 0$ in (3.5). See also Fig. 3.4a. In the following we assume that this condition holds.

Considering shared displacement directions for points on contours in 2d, *intersecting* displacement directions can cause overlap of the deformed contours, as illustrated in Fig. 3.4b. Now we examine the situation for triangular surface meshes M_1, M_2 in 3D, and displacement directions defined via a mapping m as before. For $a \in \{1, 2\}$, for surface mesh $M_a := (V_a, E_a, F_a)$ and subset of vertices U_a, we assume that a collection formed by vertices U_a, edges $\{(j, k) \in E : v_j, v_k \in U_a\}$, and triangles $\{(j, k, l) \in F : v_j, v_k, v_l \in U_a\}$ forms a (sub-) mesh (cf. Sec. 2.1.3), denoted as N_a. We assume that meshes N_1, N_2 are free of self-intersections. We assume furthermore that m maps vertices that build triangle triples in F_1 to vertices that build triangle triples in F_2. Then the mapping m induces a bijective mapping $\tilde{\mathrm{m}} : \mathfrak{N}_1 \to \mathfrak{N}_2 : \alpha v_j + \beta v_k + \gamma v_l \mapsto \alpha \mathrm{m}(v_j) + \beta \mathrm{m}(v_k) + \gamma \mathrm{m}(v_l)$. Displacing vertices

$v \in U_1$ yields displacement of all $x \in \mathfrak{N}_1$ in directions $\ell(x) := \text{sign}(x)\frac{\tilde{m}(x)-x}{|\tilde{m}(x)-x|}$.
An overlap after surface deformation cannot occur if the mapping $\tilde{m}$ satisfies the
following condition:

No two connections of two mapped point pairs intersect. (3.7)

For a proof, Let $\hat{\mathfrak{N}}_1$ and $\hat{\mathfrak{N}}_2$ be deformed surfaces. $\hat{\mathfrak{N}}_1$ is the image of a function f
that maps each point x on $\mathfrak{N}_1$ to a point $\hat{x}$ on the line $\{x + \lambda \cdot \ell(x)|\lambda \in \mathbf{R}\}$. Each
$\hat{x}$ is defined by an individual $\lambda_f(x) \in \mathbf{R}$. Likewise $\hat{\mathfrak{N}}_2$ is the image of a function g
that maps each x on $\mathfrak{N}_1$ to an $\hat{y} = x + \lambda_g \cdot \ell(x)$. If the deformed regions overlap,
i.e. intersect, there are $x_1, x_2 \in \mathfrak{N}_1$ with $f(x_1) = g(x_2)$, which means

$$x_1 + \lambda_f(x_1) \cdot \ell(x_1) = x_2 + \lambda_g(x_2) \cdot \ell(x_2) .$$

Constraint (3.6) implies $\lambda_f(x) < \lambda_g(x)$ and hence $x_1 \neq x_2$. This implies an intersection of the lines that connect the mapped point pairs $(x_1, \tilde{m}(x_1))$ and $(x_2, \tilde{m}(x_2))$,
respectively. In reverse, if no two connections of two mapped point pairs intersect,
the deformed regions do not overlap.

Construction of Shared Intensity Profiles

In the following, we propose a construction algorithm for shared displacement directions on pairs of adjacent triangular meshes, M_1 and M_2. In the process, we
partly modify the meshes' topologies (i.e. vertices and their connectivity). The one-
sided surface distance of the modified surface to the original surface is always zero.
No general assertions can be made regarding the reverse direction of the surface
distance.

Fig. 3.5 shows the construction pipeline for an exemplary femur and ilium. First,
the *mid-surface* between the objects, i.e. all points with same distance to $\mathfrak{M}_1$ and
$\mathfrak{M}_2$, is computed as the zero level of the objects' distance transforms, subtracted
from each other (Fig. 3.5b). The mid-surface is triangulated to form a mesh of
high regularity, M. Then we seek the *skeleton* of $\mathfrak{M}$. The skeleton of a surface is
the set of points that are centers of spheres that touch the surface in more than
one point, but do not intersect with the surface (see e.g. Prohaska (2007)). We
compute the skeleton of $\mathfrak{M}$ approximately by iteratively displacing the vertices of
M uniformly along their vertex normals with a small stepsize in both directions
and identifying self intersection points of the thereby displaced surface mesh in
each step. Then we identify the vertices of M whose vertex normals, scaled to a
user-specified maximal length, enter both femur and ilium with an angle bigger than
some threshold $\alpha > 0$ without reaching the skeleton of $\mathfrak{M}$ first (Fig. 3.5b and 3.5c).
This region is displaced onto each surface, $\mathfrak{M}_1$ and $\mathfrak{M}_2$, in vertex normal direction.
Each displaced patch is merged into the respective surface mesh by removing all
original triangles on the mesh which are surrounded by the patch boundary, and

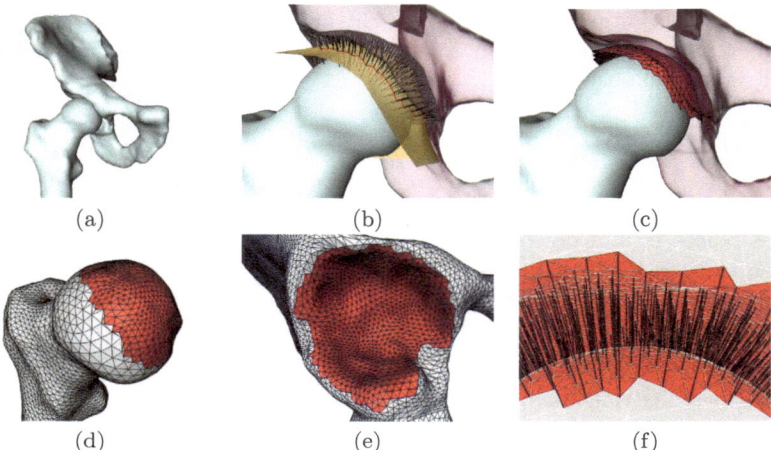

Figure 3.5 Construction of shared profiles. (a) Surface models of proximal femur and right ilium. (b) Mid surface (yellow) in region of interest, with normals entering both femur and ilium. (c) Extracted region on mid surface (red). (d,e) Surface meshes with integrated displaced region (red). (f) Vectors coupling femoral head and acetabulum.

connecting the boundaries of the remaining mesh and the patch. Fig. 3.5d and 3.5e show the resulting surfaces.

As a result, we obtain a bijective mapping of the displaced (continuous) patches that satisfies condition (3.7). Thereby shared profiles between the modified surface meshes' vertices are defined, as shown in Fig. 3.5f. We let the shared profiles reach *into* the surfaces until they meet the skeleton of $\mathfrak{M}$, or the inner skeleton of the respective surface, or until they reach a user-specified maximal length. Figure 3.6a-c shows a liver model and a model of surrounding ribs which are coupled. Three regions with shared profiles are identified here. As another example, Figure 3.6d-f shows models of a skull and a mandibular bone with shared profiles. Multiple regions are coupled here, as well.

The *mapped regions* $\mathfrak{N}_1$ and $\mathfrak{N}_2$ are guaranteed not to overlap after surface deformation along shared profiles. If initial surfaces do not overlap, deformed surfaces are guaranteed not to overlap either. An *initial overlap* of surfaces is resolved by deformation if overlapping regions can be mapped. However, an initial overlap in regions that cannot be mapped is not resolved with our method. For exemplary situations, see Fig. 3.4c. Appropriate initial segmentations are required to avoid situations as in Fig. 3.4c (bottom). See Section 8.1 for a respective discussion.

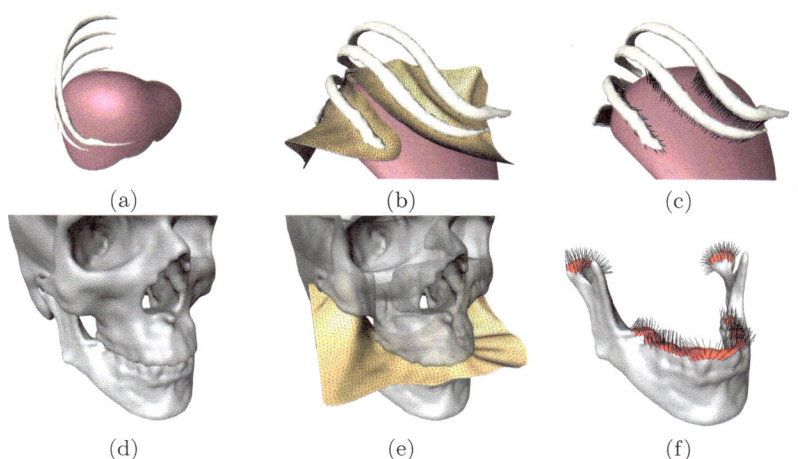

(a)   (b)   (c)

(d)   (e)   (f)

Figure 3.6 (a) Surface models of liver and nearby ribs. (b) Mid surface in region of interest. (c) Coupling vectors between liver and ribs. (d) Surface models of skull and mandibular bone. (e) Mid surface. (f) Coupling vectors between mandibular bone and skull.

3.6   Conclusion

In this chapter we have assembled a toolkit for fully automatic segmentation with deformable meshes. The toolkit can be applied for segmentation of anatomical structures whose shapes can be modeled by statistical shape models. We have presented state-of-the-art methods, as well as the following own incremental contributions: (1) An iterative method for shape-constrained free mesh deformation that prevents self-intersections of meshes and stays within a narrow band around an initial shape (cf. Sec. 3.4.1). (2) A method for constructing shared displacement directions on arbitrary adjacent surface meshes (cf. Sec. 3.5.2). This method allows for subsequent overlap-free simultaneous deformations with GraphCuts (cf. Sec. 3.5).

In Chapters 7 and 8 we will assemble fully automatic segmentation pipelines from this toolkit and apply them for segmentation of specific anatomical structures. The choice of deformation and appearance models will vary depending on the degree of shape variability and specific image appearance of a structure. We will present thorough evaluations of segmentation accuracy on clinical image data.

Besides accurate results as compared to related work, evaluation will bring to light certain limitations which are inherent to the tools described so far. In Chapter 4 we will develop a novel tool for our segmentation toolbox which overcomes these limitations.

Chapter 4

Omnidirectional Displacements for Deformable Surfaces (ODDS)

Contents

State of the art methods for mesh deformation as described in Chapter 3 search for a target surface in an image along *line segments* (typically surface normals) at each vertex of a deformable mesh (cf. Sec. 3.3). This approach has certain advantages as detailed in Section 3.3.1. However, these *unidirectional* sets of candidate locations per vertex come with the disadvantage of *restricted visibility*. In consequence, certain deformations cannot be achieved, as detailed in Section 4.1. Restricted visibility results in systematic segmentation inaccuracies.

In this chapter, we propose *omnidirectional displacements for deformable surfaces (ODDS)*, a novel deformation model which overcomes this limitation. As described in Section 4.2, ODDS allow each vertex to move not only along a line segment but within a surrounding *ball*. ODDS yield globally optimal deformations in terms of an objective function that captures appearance match and regularization penalties. As regularization penalties solely apply to local vertex neighborhoods, ODDS yield *shape-constrained free deformations* of meshes (cf. Sec. 3.4).

Allowing a ball-shaped instead of a linear range of motion per vertex significantly increases run-time and memory requirements. To alleviate this drawback of ODDS, Section 4.3 proposes a hybrid approach, *fastODDS*, with improved run-time and reduced memory requirements. Furthermore, fastODDS can also cope with simultaneous segmentation of multiple objects. The key idea of fastODDS is to employ omnidirectional displacements only in regions of high *mesh curvature*,[1] while restricting displacements to surface normals in "flat" surface regions.

An extensive evaluation of ODDS and fastODDS on clinical data is provided in Chapter 9, where we also address the influence of mesh resolution and mesh consistency. In summary, our results indicate that

1. ODDS allow for accurate segmentation of anatomical structures in regions of high surface curvature, where previous approaches based on normal displacements fail.

2. fastODDS keep all the benefits of ODDS for highly curved surface regions, while being twice as fast and requiring 50% less memory.

3. In contrast to ODDS, fastODDS can also be applied successfully for simultaneous segmentation of multiple objects.

The work presented in this chapter and Chapter 9 has been published in Kainmueller et al. (2013, 2010).

4.1   The Visibility Problem

Unidirectional sets of candidate locations have certain advantages as described in Section 3.3. Yet they suffer from *restricted visibility*: As shown in Figure 4.1, they are prone to miss features in the image data; In case of high mesh curvature

---

[1]Note that throughout this chapter, we employ the term *(mesh) curvature* to refer to the magnitude of the first principal curvature estimated on vertices of triangle meshes as proposed by Hildebandt et al. (2005). However, we *compute* mesh curvature only for the purpose of automated ridge delineation on surfaces (cf. Sec. 4.3.5).

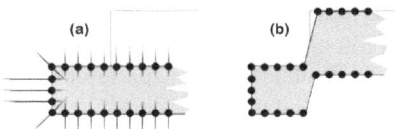

Figure 4.1 2D sketch of an exemplary deformable mesh (dark gray, with vertices as black dots) and target object (light gray). (a) Normal sets of candidate locations (directions indicated by red/gray lines) on a tip-like structure detect no target boundary points for a large set of vertices. (b) Resulting unregularized deformation onto target object boundary. Avoiding self-intersection of the mesh suppresses displacement of the bottom-left-most vertex.

unidirectional displacements may not intersect with the target boundary at all. In case *global* deformations like rigid transformations or SSM deformations are employed, this problem may be alleviated by the fact that individual ranges of motion of vertices are not tightly coupled to their respective sets of candidate locations (cf. Sec. 3.1.3). On free deformations, however, the problem has a severe impact. E.g., local translations of highly curved surface regions such as tips or ridges cannot be achieved along surface normals (cf. Figure 4.1 and Figure 4.3). This holds true independently of the chosen mesh resolution.

One heuristic approach that follows the goal of alleviating the visibility problem is repeated, i.e. *iterative*, search for appearance match and respective deformation, where the assumption is that visibility improves in the next iteration (cf. Sec. 3.4.1). There is, however, no guarantee to this end: Section 4.2.5 presents counter-examples (cf. Figure 4.3c). Furthermore, iterative deformation of meshes may easily lead to mesh inconsistencies such as self-intersections (Park et al., 2001). This requires additional remedial actions such as adaptive step-size control, adaptive re-meshing or mesh surgery (Bucki et al., 2010).

In the following section we propose a method to overcome the visibility problem for shape-constrained free deformations. The basic idea is to enlarge the set of candidate locations for which appearance match is assessed to allow not only unidirectional but *omnidirectional* displacements at each vertex of a deformable mesh.

4.2 ODDS: Free Mesh Deformations with All-around Visibility

For a more thorough search for appearance match in terms of the visibility problem (cf. Sec. 4.1), we propose to extend the set of candidate locations at each vertex of a deformable surface mesh from a line segment to a ball centered at the respective vertex position. Thus we achieve all-around visibility within some radius. Given such ball-shaped sets of candidate locations, we formulate the mesh deformation problem as a trade-off between finding good appearance match within these balls and simultaneously considering local regularization.

Volumetric (three-dimensional) ball-shaped sets of candidate locations of neighboring vertices overlap heavily in case the ball radius is bigger than the distance between the respective vertices; Furthermore, individual sets of candidate locations most probably contain a whole region (two-dimensional manifold) of the target surface. Hence highly inconsistent (dissimilar) displacements on neighboring vertices may point to the target surface. The type of local regularization we employ must be able to avoid inconsistent displacements of adjacent vertices.

To this end, we asses appearance match at – and allow displacements to – a *discrete* set of points within a ball around each vertex (Sec. 4.2.1). Free deformations are modeled by penalizing differences of displacements of edge-connected mesh vertices (Sec. 4.2.2). This discrete setting enables us to frame the mesh deformation problem as the problem of minimizing the energy of a *Markov Random Field* (MRF, see e.g. Koller and Friedman (2009), Sec. 4.2.3). MRFs can be optimized efficiently (Komodakis et al., 2008), which has made them attractive for many applications in image processing and computer graphics (see e.g. Glocker et al. (2008); Paulsen et al. (2010); Li (2009); Blake et al. (2011)). We denote the method of ball-shaped sets of candidate locations combined with MRF optimization for surface mesh deformation as *omnidirectional displacements for deformable surfaces*, or *ODDS*.

In Section 4.2.5, we demonstrate the theoretical benefits of ODDS with experiments on synthetic data.

4.2.1 Omnidirectional Displacements

We define candidate displacements $s \in S \subset \mathbf{R}^3$ to be distributed within a ball of radius r_S, i.e. $\forall s \in S : \|s\| < r_S$, where r_S is a parameter of our method. We arrange displacements in S are arranged as a cubic close-packed lattice (Conway et al., 1999); see Fig. 4.2a for a 2D sketch. We refer to such a ball-shaped set of displacements as *omnidirectional displacements*.

Note a conceptual difference between omnidirectional displacements and conventional, unidirectional displacements (cf. Sec. 3.3): Omnidirectional displacements are not defined per individual vertex, but are equal for all vertices of a mesh. In consequence, applying the "same" displacement to different vertices means shifting these vertices by the same, three-dimensional vector (see Fig. 4.2b,c). In contrast, unidirectional displacements are specified via a *set of lengths* which is equal for all vertices, yet *vertex-individual* sets of displacements are obtained as vectors of the respective lengths in individual directions; In other words, "same" displacements

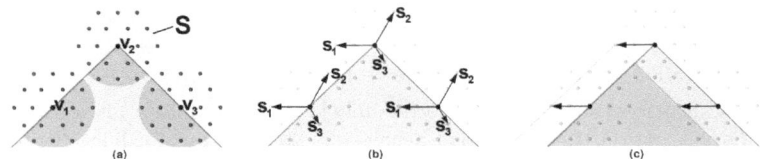

Figure 4.2 2D sketch of omnidirectional displacements: (a) Black dots depict three vertices v_1, v_2, v_3 of a deformable mesh. Ball-shaped ranges of motion S (large gray disks) around each vertex are discretized via sample points (yellow/gray dots). (b) Exemplary displacements s_1, s_2, s_3 to sample points are shown as black arrows, where equivalent displacements on different vertices are indicated by corresponding numbers. (c) Applying the same displacement to all vertices leads to parallel translation.

on different vertices have same lengths but may have different directions.

4.2.2   The Mesh Deformation Problem

As before (cf. Sec. 2.2, 3.3, 3.2.1), for each displacement $s \in S$ and vertex $v \in V$, a scalar cost $\phi(v, s) \geq 0$ encodes whether sample point $v + s$ is believed to lie on the target object boundary within the image $I : \Omega \subset \mathbf{R}^3 \to \mathbf{R}$. The stronger the belief, the lower the cost. In other words, $\phi(v, s)$ serves as a penalty for the case that v is displaced by s. In general, however, any $\phi : V \times S \to \mathbf{R}_0^+$ is feasible as appearance cost function.

For each two displacements s_i, s_j, a scalar cost $\psi(s_i, s_j) \geq 0$ serves as a penalty for the case that s_i and s_j occur on adjacent vertices. The function $\psi : S \times S \to \mathbf{R}_0^+$ takes care of regularization. In the following, we assume that ψ is monotonically increasing with the Euclidean norm of the difference of displacements, $\|s_j - s_i\|$, and depends on nothing else. Whenever it adds to clarity, we sloppily denote $\psi(\|s_j - s_i\|) := \psi(s_i, s_j)$.

We formulate the mesh adaptation problem as the following optimization problem:

$$\text{Find } \{d_i \in S\}_{i=1}^{n_V} \text{ for which } \left(\sum_{i=1}^{n_V} \phi(v_i, d_i) + \sum_{(j,k) \in E} \psi(d_j, d_k) \right) \text{ is minimal.} \quad (4.1)$$

This means we are looking for a displacement field d that minimizes an objective function that sums up the appearance costs and regularization penalties for all vertices. This problem formulation implies that the set $v + S$ defines the set of candidate locations as well as the range of motion for vertex v. Hence effective resulting displacements are elements of S.

Note that (4.1) contains an implicit parameter that controls the trade-off between "image fit" and regularization. It can be adjusted by scaling the cost- or the distance function. Section 4.2.5 describes how we perform this scaling.

Sets of displacements S are not vertex-individual but equal for all vertices of a mesh (cf. Sec. 4.2.1). Regularization penalties proportional to Euclidean distances of displacements have the effect of penalizing local scaling of the mesh (i.e. growing or shrinking), while parallel translations are not penalized (see Fig. 4.2c). We consider this beneficial as we expect our initial meshes as well as their local features to have approximately correct scale. Alternatively, if scaling should not be penalized, one could penalize differences between *parameter vectors* that describe displacements w.r.t. local, vertex-individual coordinate systems. As with unidirectional displacements, one local coordinate axis could be the respective vertex normal.

4.2.3 Optimal Mesh Deformation via MRF Energy Minimization

The objective in (4.1) has the form of a *first-order MRF energy* (see e.g. Li (2009)), with vertices being represented by MRF-nodes, mesh edges by MRF-edges, and displacements by the possible states (also called *labels*) of the nodes. Cost $\phi(v, s)$ defines the unary potential of node v in state s, and penalty $\psi(s_i, s_j)$ defines the binary potential of two adjacent nodes in states s_i, s_j. The MRF-state with minimal sum of potentials yields the desired displacement field. We optimize the MRF energy with a solver proposed by Komodakis et al. (2008). This solver is guaranteed to find an approximately optimal solution. Furthermore, it can deal with non-metric distance functions ψ. It solely requires ψ to satisfy $\psi(s_i, s_j) = 0 \Leftrightarrow s_i = s_j$.

4.2.4 Refined Regularization

The condition $\psi(s_1, s_2) = 0 \Leftrightarrow s_1 = s_2$ has the effect that there is always a penalty for unequal displacements on neighboring vertices. In other words, even the smallest distance between displacements, i.e. the sampling distance δ_S, is penalized if respective displacements occur on neighboring vertices. The sampling distance serves as a scale on which features shall be detected in the image data; in general it is not supposed to determine the amount of regularization imposed upon mesh deformation. A straightforward option to "tolerate" some larger distance between displacements while respecting the condition $\psi(s_i, s_j) = 0 \Leftrightarrow s_i = s_j$ would be to set the respective penalties to a very small value with respect to all others. However, setting very small binary potentials to obtain "almost" unpenalized displacement distances impairs the approximate optimality guarantees of the MRF solver (Komodakis et al., 2008), which depend on the ratio between largest and smallest non-zero binary potential.

Alternatively, we propose to approximate a "tolerated distance" with zero penalty as follows: Let $\tilde{S} := \{\tilde{s}_i \in \mathbf{R}^3 \mid i = 1 \dots n_{\tilde{S}}\}$ be a second cubic close-packed lattice which is coarser than S, i.e. $\delta_{\tilde{S}} > \delta_S$. $\tilde{S}$ partitions S into *displacement blocks* B_i by means of nearest-neighborhood to its elements $\tilde{s}_i$. Formally,

$$B_i := \left\{ s \in S \middle| \tilde{s}_i = \operatorname*{argmin}_{\tilde{s} \in \tilde{S}} \|s - \tilde{s}\| \right\} .$$

Given these displacement blocks, we set up an MRF with states $\tilde{s}_i$ via unary potentials

$$\tilde{\phi}(v, \tilde{s}_i) := \min_{s \in B_i} \phi(v, s) ,$$

and binary potentials

$$\tilde{\psi}(\tilde{s}_i, \tilde{s}_j) := \psi(\|\tilde{s}_i - \tilde{s}_j\|) .$$

We compute $\tilde{d} : V \to \tilde{S}$ which minimizes the respective MRF energy. We assign to vertex v_k with $\tilde{d}(v_k) = \tilde{s}_i$ the displacement $s \in B_i$ with minimum cost, i.e. $d_k := \operatorname{argmin}_{s \in B_i} \phi(v_k, s)$.

The sampling distance of $\tilde{S}$ defines an upper bound to the Euclidean norm of displacement differences that are not penalized. More precisely, with block sampling distance $\delta_{\tilde{S}}$, zero penalty is attributed to displacements with $\|s_i - s_j\| < \delta_{\tilde{S}}$ in case s_i and s_j belong to the same block, while the minimum non-zero penalty is attributed to displacements with $\|s_i - s_j\| < 2\delta_{\tilde{S}}$ in case s_i and s_j belong to adjacent blocks.

Note that the proposed approach allows for the "best" approximative optimality guarantee of the MRF solver (Komodakis et al., 2008) given the desired amount of regularization. Furthermore, memory requirements are significantly reduced as compared to the straightforward approach: The size of the MRF to be solved depends only on the desired amount of regularization, and does not increase in case of refined sampling of candidate displacements. These advantages come at the cost of distance penalties not only depending on actual displacement distances, but also on displacement block organization.

4.2.5 Proof of Concept Synthetic Experiments

For a proof-of-concept qualitative evaluation, we applied ODDS to (1) Synthetic binary images, and (2) synthetic binary images with various amount of noise. For comparison, we also computed results with FreeBand and GraphCuts (cf. Section 3.4).

We computed appearance costs $\phi(v, s) \in \mathbf{R}$ as proposed in Sec. 3.2.1, with parameters set to $t_1 := 0.1$, $t_2 := 1.1$, and $g := 0.1$. Note that for the computation of costs at a vertex v, image derivatives are assessed in the *same* direction for all displacements s, namely in direction of the surface normal n_v at v, no matter which type of displacements are employed (unidirectional or omnidirectional).

	$\#V$	$\bar{e}$	$2r$	δ_S	$\#S$	$\#\tilde{S}$
Cube	770	1	26	0.5	105294	3768
Ellipsoid	1797	1	31	0.5	178201	6989

Table 4.1 Application specific parameters are the number of vertices $\#V$ of the deformable mesh (which determines the average edge length of mesh triangles, $\bar{e}$ [mm]), the diameter $2r$ [mm] and sampling distance δ_S [mm] of sets of displacements, the number of sample points $\#S$, and the number of MRF labels $\#\tilde{S}$ employed for ODDS.

As for the trade-off between appearance match and regularization, we scale the appearance cost function such that $a \cdot \psi(\delta_{\tilde{S}}) < (\phi_{high} - \phi(v, s)) < a \cdot \psi(2\delta_{\tilde{S}})$ for any cost $\phi(v, s) < \phi_{high}$, where $a = 6$ is the average number of edges per vertex. Image derivatives are scaled such that $0 < \phi(v, s) < 0.5 \cdot \phi_{high}$ for any $\phi(v, s) < \phi_{high}$. This serves for a clear distinction of matching from "non-matching" appearance, but also leaves room to distinguish between the quality of features, i.e. there exist $\phi_1 < \phi_2 < \phi_{high}$ with $a \cdot \psi(\delta_{\tilde{S}}) < (\phi_2 - \phi_1)$.

Whenever we compare different adaptation methods on the same image data, we use the same appearance cost function ϕ for all methods.

For omnidirectional displacements, we employ displacement blocks with sampling distance $\delta_{\tilde{S}} = 3\delta_S$; as distance function ψ, we use $\psi(s_1, s_2) = \|(s_2 - s_1)/\delta_{\tilde{S}}\|^3$ in all experiments. The GraphCuts regularization parameter c equals the block sampling distance $\delta_{\tilde{S}}$ as set in the respective ODDS experiment, i.e. $c = \delta_{\tilde{S}}$;

Whenever we employ unidirectional as well as omnidirectional displacements for the same image data (in multiple methods), the "radius" of the unidirectional range of motion equals the respective ball radius, i.e. $r_L = r_S$. As for the sampling distance of unidirectional displacements, we set it to half the sampling distance of the respective omnidirectional displacements, i.e. $\delta_L = 0.5\delta_S$. In contrast to GraphCuts and ODDS, FreeBand adaptations were performed iteratively, with 30 steps.

Table 4.1 lists the values of experiment-specific parameters.

Results

We performed experiments on binary images[2] of a cube and a thin ellipsoid. As initial meshes, we used triangulated cube and tip surfaces with ideal shape, but shifted pose (see Fig. 4.3a). We chose ball radii such that the target object boundary was located completely within a band of respective width around the initial mesh. Fig. 4.3b shows the results of adding normal displacements without any

---

[2]i.e. intensities $\in \{0, 1\}$

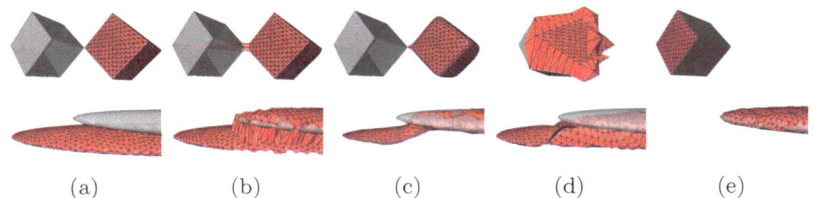

(a)   (b)   (c)   (d)   (e)

Figure 4.3 Results on synthetic data. Deformable mesh (red) and target object (transparent gray surface) are shown (a) in their initial situation, and after deformation via (b) displacements along normals without regularization, (c) FreeBand, (d) GraphCuts, and (e) ODDS.

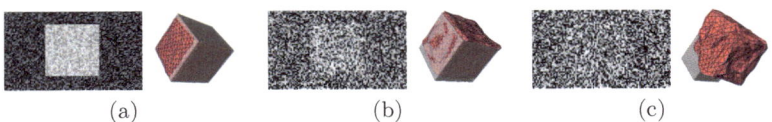

(a)   (b)   (c)

Figure 4.4 Performance of ODDS in the presence of noise. We added random noise with range (a) $[-0.5, 0.5]$, (b) $[-2.5, 2.5]$ and (c) $[-5, 5]$ to a binary image of a cube. We show slices of the image data and the respective adaptation result (red mesh). The gray transparent surface depicts the ideal target object.

regularization. The results of FreeBand-, GraphCuts- and ODDS adaptation are shown in Fig. 4.3c, 4.3d and 4.3e, respectively.

We added various amounts of random noise to the binary cube image and performed ODDS as before. The cube was detected correctly for noise with ranges $[-0.5, 0.5]$ and $[-2.5, 2.5]$, and failed for $[-5, 5]$. Fig. 4.4 shows slices of the noisy image data and the respective adaptation results.

Discussion

Experiments on synthetic binary images show that ODDS are able to handle parallel translations of highly curved surface regions, in contrast to conventional free deformation approaches (GraphCuts and FreeBand) that employ normal displacements. While global deformation models (like e.g. rigid transformations) may also yield the desired vertex displacements in the experiments we present, ODDS achieve them by means of a *free* deformation model.

Experiments on noisy synthetic images show that ODDS are able to produce well-regularized displacement fields in the presence of noise. However, for a very low signal-to-noise ratio, ODDS may fail to detect the target object.

An in-depth quantitative evaluation of ODDS on clinical data is presented in Chapter 9.

4.3 FastODDS

ODDS are designed to allow for accurate segmentations of highly curved structures, where methods that employ unidirectional displacements are fundamentally limited (cf. Fig. 4.1). The methodological benefits of ODDS come with drawbacks in computational efficiency: Allowing a three-dimensional set of candidate locations per mesh vertex significantly increases run-time and memory requirements as compared to unidirectional search spaces. The required number of sample points per vertex for a ball-shaped range of motion with radius r is in $O(r^3)$, while it is in $O(r)$ for line segments. Run-time and memory requirements for computing appearance costs behave accordingly.

Overall memory requirements for running ODDS on a mesh (V, E, F) are in $O(r^3(|V| + |E|))$, as opposed to $O(r(|V| + |E|))$ for GraphCuts. Note that the overall run-time of ODDS on a mesh is dominated by the computation of ϕ (cf. Sec. 9.2.3).

For instance, given a medium-sized mesh with about 8500 vertices, together with a displacement set S with diameter 15 mm and sampling distance 0.4 mm, ODDS take about 2:30 minutes to compute on a 3.5 GHz core and require more than 4.5 GB of memory, while GraphCuts take about three seconds and require less than one GB of memory (cf. Tab. 9.5).

Apart from the above-mentioned limitations, unidirectional displacements *do* allow for an accurate segmentation of "flat" structures, where the visibility problem is at least unlikely to occur. Anatomical structures of interest in medical image analysis often exhibit mainly flat or only slightly curved surface regions, while high curvature appears on a much smaller amount of their surface.

Therefore we propose to use omnidirectional displacements only in (and next to) surface regions with high curvature, while employing unidirectional displacements in flat surface regions (see Sec. 4.3.1). Thus we exploit the benefits of ODDS, while reducing run-time and memory via an overall reduced amount of sample points. We call this approach *fastODDS*.

We propose to compute unidirectional and omnidirectional displacement sequentially (see Sec. 4.3.3). Hence, in general, any method for obtaining unidirectional displacements can be chosen. In this work, we employ the method of Li et al. (2006), which we refer to as GraphCuts (see Sec. 3.4.2). As described in Section 4.3.4, the multi-object ability of GraphCuts (cf. Sec. 3.5) can be transferred to fastODDS.

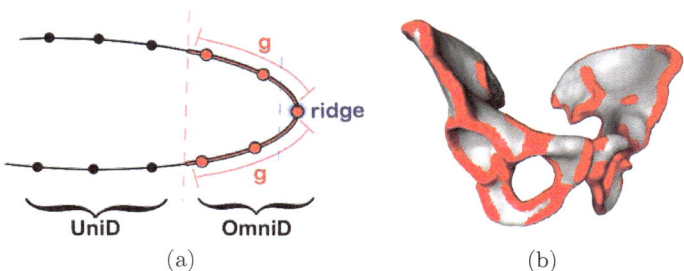

(a)   (b)

Figure 4.5 (a) 2D-sketch of OmniD- and UniD region on an exemplary tip-like mesh (vertices depicted as dots) with *ridge* at rightmost vertex (blue dot). Only a small region around the ridge exhibits high curvature, as indicated by the light blue, dashed line. Instead, we assign vertices within a certain *geodesic distance* g around the ridge to the OmniD region; all others belong to the UniD region. (b) Exemplary hip bones with OmniD region (red) and UniD region (gray).

4.3.1 Where to use Omnidirectional Displacements

When defining the surface region where omnidirectional displacements shall be applied (OmniD region), we assume that we want to achieve a smooth transition to the region where unidirectional displacements shall be applied (UniD region). Consider e.g. a sharp ridge surrounded by flat surface regions. Imagine we want to translate this ridge in a direction roughly parallel to the flat surface regions. To achieve a smooth overall displacement field, surface-tangential movements cannot be allowed on the boundary of the OmniD region. Therefore we need to employ omnidirectional displacements not only in the region of high curvature (i.e. on the ridge and in a very small area around it), but within a larger *transition region* around the ridge. Hence, a band of some width around high-curvature regions has to be included in the OmniD region (see Fig. 4.5a).

We propose to define the OmniD region as follows: (1) Identify *ridges* on the surface. Ridges may be computed automatically on an initial segmentation (see Section 4.3.5), or, in case a statistical shape model is used for initial segmentation, defined a priori (automatically or manually) on the model. (2) Identify the surface region that lies within a certain *geodesic distance* g to those ridges. As for the UniD region, we define it as the complement of the OmniD region on the surface. Fig. 4.5b shows OmniD- and UniD regions on an exemplary anatomical structure.

The geodesic distance threshold g is a parameter of fastODDS. Informally speaking, it should be large enough to allow for the desired amount of displacement of ridge vertices without too much regularization penalty. Consider a deformable mesh with mean edge length $\bar{e}$. Then, $g/\bar{e}$ roughly estimates the number of edges that connect a ridge to the boundary of the OmniD region. Stretching (or shrinking) each of these edges by one sampling distance δ can reach a translation of the ridge up to $\delta \cdot g/\bar{e}$. In case we can estimate a desired maximum amount of displacement $t \in \mathbf{R}$, we may define $g = \bar{e} \cdot t/\delta$. This way, the desired displacement of ridge vertices can be achieved with no more than the minimum non-zero penalty $\psi(\delta)$ at any edge.

4.3.2   The Hybrid Mesh Deformation Problem

We propose to compute displacements for OmniD- and UniD region with ODDS and GraphCuts, respectively. If not mentioned otherwise, we use the same notation as in Section 4.2. Let V_U be the vertices in the UniD region, V_O the ones in the OmniD region, with $V = V_U \cup V_O$ and $V_U \cap V_O = \emptyset$. The pairs of adjacent vertices E are partitioned into $E_O = (V_O \times V_O) \cap E$, $E_U = (V_U \times V_U) \cap E$ and $E_\partial = (V_O \times V_U) \cap E$. This means E_O contains the edges in the OmniD region, E_U the edges in the UniD region, and E_δ the edges that bridge between V_O and V_U. We refer to the set of vertices in the UniD region which are part of an edge that bridges to the OmniD region as *UniD boundary* $\partial V_U = \{v_k \in V_U : \exists v_j \, in V_O : (j,k) \in E_\partial\}$.

We are dealing with two sets of displacements, namely the discretized ball-shaped range of motion S which applies to all vertices in the OmniD region, and discretized linear ranges of motion $L(v)$ along directions $\ell(v)$ per vertex v of the UniD region. We assume for the moment that $\forall v \in V_U : L(v) \subset S$ (see Sec. 4.3.3). Thus we can conveniently denote a displacement field as

$$d : V \to S, v \mapsto d(v) \begin{cases} \in S : v \in V_O \\ \in L(v) : v \in V_U \end{cases} \tag{4.2}$$

Then the goal of fastODDS is to

$$\text{Find } \{d_i\}_{i=1}^{n_V} \text{ for which } \sum_{i \in N_V} \phi(v_i, d_i) + \sum_{(j,k) \in E_O \cup E_\partial} \psi(d_j, d_k) \text{ is minimal}$$

$$\text{and } \quad \forall (j,k) \in E_U : |\ell_j \cdot d_j - \ell_k \cdot d_k| \leq c \tag{4.3}$$

Note that (4.3) adds together (4.1) on the OmniD region, (3.4) on the UniD region, and regularization penalties for the edges bridging between OmniD- and UniD region.

4.3.3   Optimal Hybrid Mesh Deformation

We follow the simple idea to compute unidirectional and omnidirectional displacements sequentially, in a way that a smooth transition between UniD- and OmniD region is achieved. Therefore, we first obtain a displacement field for the UniD region (via GraphCuts or any other method), and second perform ODDS on the OmniD region, constrained by fixed displacements on the UniD boundary as computed beforehand.

This approach partitions optimization problem (4.3) into two parts that are subsequently solved. First, we solve

$$\text{Find } d|_{V_U} := \{d_i\}_{v_i \in V_U} \text{ for which } \sum_{v_i \in V_U} \phi(v_i, \hat{d}_i) \text{ is minimal}$$
$$\text{subject to } \forall (j,k) \in E_U : |\ell_j \cdot \hat{d}_j - \ell_k \cdot \hat{d}_k| \leq c \tag{4.4}$$

via GraphCuts. Second, we approximately solve

$$\text{Find } d|_{V_O} := \{d_i\}_{v_i \in V_O} \text{ for which}$$
$$\sum_{v_i \in V_O} \phi(v_i, \hat{d}_i) + \sum_{(j,k) \in E_O} \psi(\hat{d}_j, \hat{d}_k) + \sum_{(j,k) \in E_\partial} \psi(\hat{d}_j, d|_{V_U}(v_k)) \tag{4.5}$$
$$\text{is minimal,}$$

i.e. we compute a $d|_{V_O}$ which approximately minimizes (4.5), via MRF optimization. Note that in (4.5), $d|_{V_U}(v_k)$ is fixed for all $v_k \in V_U$.

While GraphCuts find a displacement field that is a global minimizer of (4.4), and MRF optimization guarantees an approximately optimal solution of (4.5) within provable bounds (Komodakis et al., 2008), our hybrid approach for solving (4.3) does not guarantee either of these global properties.

The optimality bounds which are guaranteed for MRF optimization would determine bounds for the overall objective function in case we solved (4.3) directly. A naive approach would be the following: For all feasible displacement fields on the UniD boundary, try to solve (4.4) with respective fixed displacements on the UniD boundary; In case a feasible solution exists on the UniD region, solve (4.5), as usual with respective fixed displacements on the UniD boundary; Choose the pair of solutions with minimal sum of costs. However, the number of feasible displacement fields on the UniD boundary is in $O(n_L \cdot (2c/\delta_L + 1)^{|\partial V_U|})$, and hence for $c > 0$ this approach is not feasible for performance reasons. Investigating whether there is a more efficient algorithm for solving (4.3) with the same optimality guarantees as for MRF optimization is subject to future work.

Practically, to get a "good" solution on the UniD boundary, we perform Graph-Cuts on the whole surface except vertices on or next to ridges. The overlap with the OmniD region serves for an extended regularization in GraphCuts optimization. The resulting displacements are discarded afterwards. Note that "cutting the surface open" along ridges allows for translational movements of surface regions near ridges with GraphCuts at least in one (surface-normal) direction. Otherwise, moving the surface "inward" on one side of a ridge and "outward" on the opposite side may not be possible due to the shape constraint.

As for the OmniD region, practically, we enlarge it by the UniD boundary, and achieve fixed displacements for each boundary vertex $w \in \partial V_U$ by assigning zero appearance cost to $s = d|_U(w)$ and infinite cost to all other displacements in S. More precisely, as unidirectional displacements are generally *not* in S due to discretization, we assign zero cost to the *closest* displacement in S, and infinite cost to all others:

$$\phi(w, s) := \begin{cases} 0 : s = \operatorname{argmini}_{\hat{s} \in S} \|d|_{V_U}(w) - \hat{s}\| \\ \infty \text{ otherwise} \end{cases}$$

4.3.4 Multi-object FastODDS

GraphCuts can be used for simultaneous segmentation of multiple objects via shared displacement directions for arbitrary adjacent structures (cf. Section 3.5). Hard constraints on the distance between adjacent surfaces can be enforced. To transfer this capability to fastODDS, we use multiple surfaces that are coupled with shared displacement directions in adjacent regions as input, and partition each surface into OmniD- and UniD region as for single-object fastODDS. Then, we apply multi-object GraphCuts on the (coupled) UniD regions. Subsequently, we apply ODDS on the OmniD regions as for single-object fastODDS (cf. Eq. 4.5), i.e. constrained by fixed displacements on the UniD boundary as computed beforehand via multi-object GraphCuts. This way, fastODDS can handle multi-object situations in case adjacent surface regions are, at least to some extent, flat, and hence equipped with linear range of motion.

If, however, the coupled region exhibits high curvature, it may overlap with the OmniD region. Consequently, the resulting deformed surface may intersect with the adjacent surface. This can be prevented in case we know beforehand that one of the adjacent surfaces does not exhibit high curvature. In this case, the multi-object GraphCuts result on the "flat" surface can be used to modify the cost function on the OmniD region of the "curved" surface such that no overlap can happen. This can be achieved by setting costs to infinite for all sample points that lie inside the deformed "flat" surface. However, in case both adjacent surfaces exhibit high curvature within the coupled region, multi-object fastODDS do not guarantee non-overlapping results.

4.3.5 Appendix: Automatic Ridge Detection

This section briefly introduces a method for automatic ridge detection as proposed and described in detail by Weber (2008). For ridge detection on surfaces, Weber (2008) utilizes the ridge definition first introduced by Rothe (1915), more recently described by Koenderink and van Doorn (1993). Intuitively a ridge of a *height function* can be imagined as a way one would take when walking up a mountain. One usually chooses the path with the lowest slope since it is the least exhausting.

Weber (2008) applies this definition for ridges to the maximum principal curvature $\kappa := \max(|\kappa_1|, |\kappa_2|)$ on triangle surfaces (cf. Hildebandt et al. (2005)) as height function, yielding ridges along sharp edges as well as sharp wrinkles of a surface.

Conventionally, to find ridges in terms of the above ridge definition, the fifth derivative of the surface is computed, which is very sensitive to noise (Weber, 2008). Weber (2008) uses a more robust property of these ridges which we just describe intuitively here. (Weber (2008) gives both the following intuitive as well as a formal description.) Suppose we descend along a ridge for a fixed distance f, starting at a certain height h. If, instead, we do not start at the ridge, but on the isoline of height h a little to the right or to the left of the ridge, and walk the same distance in direction of steepest descent, we will end up lower. Consequently also the *integral* of the heights we pass when starting on the ridge, H, is higher than the integrals when starting beneath the ridge, H_{left} and H_{right}, and the same holds for the average height of the walk, $\bar{h} = H/f$. Weber (2008) approximates the respective *integral curves* of the (discrete) gradient vector field of κ on a triangle surface mesh (Forman, 1998) as described by Cazals et al. (2003).

The walking distance f, i.e. the arc length for integration, is called *filter length*, specified by a number of edges n_f in our discrete setting. The difference of the average heights, $\min\{\bar{h} - \bar{h}_{left}, \bar{h} - \bar{h}_{right}\}$ [mm^{-1}], is called *significance* of a ridge. If significance is low, the ridge might not be sharply peaked. Weber (2008) suggests to discard ridge pieces if significance does not exceed a user given *significance threshold*, and also if their average height $\bar{h}$ is below a user defined *curvature threshold* [mm^{-1}], in which case ridge pieces are not necessarily strong features of the surface.

4.4   Conclusion

We proposed ODDS, a method that allows omnidirectional displacements for all vertices of a surface mesh during deformable model adaptation. We encode the mesh adaptation problem as a Markov Random Field energy which is a linear combination of vertex-wise appearance costs and regularizing penalties that affect local vertex neighborhoods. We employ an MRF solver as proposed by (Komodakis et al., 2008) to compute mesh deformations that approximately minimize MRF energy globally.

We presented proof-of-concept experiments on synthetic data where ODDS outperform traditional mesh adaptation along line segments (e.g. surface normals) in regions with high curvature (tips and ridges) in terms of segmentation accuracy.

To save run-time and memory as required by ODDS, we developed a hybrid approach, fastODDS. Here, we employ omnidirectional displacements adaptively, i.e. only where high curvature calls for them, and traditional unidirectional displacements elsewhere. Apart from computational efficiency, an additional benefit of fastODDS is that it can be applied for simultaneous adaptation of multiple, adjacent meshes, i.e. multi-object segmentation.

A thorough quantitative evaluation of ODDS and fastODDS on clinical data will be presented in Chapter 9.

Chapter 5

From Surface Mesh Deformations to Volume Deformations

Contents

SSMs as described in Section 3.1 are linear models. They are not suitable for modeling large non-linear shape variations, nor changes in shape topology. Building non-linear statistical shape models is an active area of research (see e.g. Gerber et al. (2009)). However, certain anatomical structures exhibit large non-linear *intra-individual* variations of shape that are of *biomechanical* nature rather than stemming from a statistical distribution. E.g., articulated bone compounds are subject to intra-individual shape variation stemming from articulation in joints. Such variations call for mechanical rather than statistical modeling (Kainmueller et al., 2009b).

Assuming approximately linear *inter-individual* shape variation, building an SSM for such a structure requires "alignment" of training shapes w.r.t. intra-individual variations, and furthermore incorporation of this variation into the SSM. This has been done for the case that intra-individual variations can be described by simple mechanical models, namely for the creation of articulated multi-bone SSMs (Kainmueller et al., 2009b; Bindernagel et al., 2011; Bindernagel, 2013).

As for a more complex case, e.g. soft-tissue structures of the musculoskeletal system are subject to large non-linear intra-individual shape variations stemming from bone articulation, tension and training. To the best of our knowledge, there is no simple, few-parameter mechanical model to describe such variations. In consequence, linear SSMs with additional parameters as in (Kainmueller et al., 2009b) are not straightforwardly suitable for modeling respective structures. Direct modeling via non-linear statistical shape models disregards the discrimination between (large) intra-individual, biomechanical variations on the one hand and inter-individual, statistical variation on the other hand. We hypothesize that large intra-individual variations obstruct an effective learning of inter-individual variations with non-linear statistical shape models.

Instead, this chapter describes alternative methods for segmentation of anatomical structures which do not require modeling their shapes by means of statistical shape models. We focus on the case that *nearby* structures *can* be modeled and segmented successfully via SSMs. This case holds e.g. for soft-tissue structures of the musculoskeletal system which are located in the vicinity of articulated bones.

In Section 5.1 we describe methods for extrapolating surface mesh deformations to deformations of the embedding space $\mathbf{R}^3$. This form of extrapolation opens a way for segmentation of structures not suitable for modeling by SSMs but "adjacent" to SSM-modellable structures. Segmentation by extrapolation can be refined by *atlas-based segmentation*, which we describe in Section 5.2. In Chapter 10 we apply the methods described in this chapter for segmentation of individual leg muscles in CT, and present a quantitative evaluation of segmentation accuracy on clinical data.

5.1   Mesh-based Extrapolation

5.1.1   Introduction

Let $M := (V, E, F)$ be a surface mesh that is deformed by means of vertex displacements $v \mapsto \hat{v}$, yielding a deformed surface mesh $\hat{M}$. Hence M and $\hat{M}$ have the same mesh structure; they only differ in terms of vertex positions. We assume intersection-free meshes in the following. Then, a respective *surface deformation* that maps all points on $\mathfrak{M}$ by means of their barycentric coordinates (cf. Sec. 2.2) can be written as

$$\mathrm{m} : \mathfrak{M} \to \hat{\mathfrak{M}}, \alpha v_j + \beta v_k + \gamma v_l \mapsto \alpha \hat{v}_j + \beta \hat{v}_k + \gamma \hat{v}_l \tag{5.1}$$

A method for extrapolation of surface mesh deformations onto $\mathbf{R}^3$ is a function extra that maps m to extra(m) := e : $\mathbf{R}^3 \to \mathbf{R}^3$.

Extrapolation of surface mesh deformations allows for estimating segmentations of structures not suitable for modeling by SSMs but located in the vicinity of SSM-modellable structures: Segmenting an SSM-modellable structure by means of SSM deformation yields mappings of each *training mesh* $M^{(k)} := (V^{(k)}, E, F)$ used for SSM generation to the resulting deformed mesh $M^* := (V(b^*, T^*), E, F)$, $m^{(k)} : \mathfrak{M}^{(k)} \to \mathfrak{M}^*$. Any such surface deformation $m^{(k)}$ can be extrapolated onto structures depicted in the respective training image. In case a reference segmentation of such a structure is given, applying an extrapolated deformation to the reference segmentation yields a respective (estimated) segmentation of the target image. Such an estimated segmentation can be refined by means of image-to-image registration, and merged with segmentations obtained via other training shapes, as described in Section 5.2. This way, SSM-modellable structures can be exploited for segmentation of (adjacent) structures that are not suitable for modeling with SSMs.

Furthermore, extrapolation as described in this section can also be used to *refine an SSM-based extrapolation* of a structure for which an appearance model is not available (cf. Sec. 3.1.5): Let V denote the vertices of an SSM instance (b, T). In case V is determined by SSM deformation via an appearance model of a *subset* of vertices $W := \{v_i(b, T) | i \in \mathcal{J} \subset N_V\}$, vertices $U := V \backslash W$ are extrapolated by means of SSM shape parameters. A function extra can extrapolate a subsequent free deformation of W, $\mathfrak{W} \to \hat{\mathfrak{W}}$, yielding a free deformation of U, $\mathfrak{U} \to \hat{\mathfrak{U}}$ without the need to establish an appearance model for U.

Properties of Extrapolated Deformations

Extrapolation of surface deformations onto volumes is intended to yield "anatomically plausible" deformations of tissues. Ideally, resulting volume deformations should mimic intra-individual, physical behavior of tissue, e.g. subject to bone articulation, and add to that plausible inter-individual deformations as far as they can be deduced from respective surface deformations.

Mathematically, extrapolated deformations should be *exact* on $\mathfrak{M}$, i.e. an extrapolation method should satisfy $\forall m \, \forall x \in \mathfrak{M} : (\mathrm{extra}(m))(x) = m(x)$. Furthermore, based on the assumption that anatomically plausible deformations cannot *fold* tissue, an invertible m should yield an invertible e. Another property that we assume to hold for anatomically plausible deformations is *differentiability* at least within homogeneous tissues. To this end, note that a surface mesh deformation m as defined by (5.1) is continuous but ·not differentiable on $\mathfrak{M}$. Consequently, an exact extrapolation method can only yield a deformation e which is differentiable at most on $\mathbf{R}^3 \backslash \mathfrak{M}$.

In the following, we describe three extrapolation methods, each satisfying some of the above conditions. The methods we discuss perform extrapolation *geometrically*: They operate solely on the basis of m and do not incorporate any further prior knowledge. Geometric extrapolation methods are conceptually simple, computationally efficient and yield closed-form solutions. However, to the best of our knowledge, there is no such method that satisfies *all* of the above properties. Analyzing alternative extrapolation methods which explicitly model physical properties of tissue (see e.g. Duysak et al. (2003); Klapproth et al. (2012)) is subject to future work.

Section 5.1.2 describes simple extrapolation by means of a single affine transformation. Section 5.1.3 describes *polyaffine transformations*, a non-linear method proposed by Arsigny et al. (2005) which yields differentiable and invertible extrapolated deformations, yet is not exact. Section 5.1.4 describes a method proposed by Ju et al. (2005) which is exact and extrapolated deformations e are differentiable on $\mathbf{R}^3 \backslash \mathfrak{M}$ yet in general not invertible.

5.1.2 Affine Transformations

Given a surface mesh deformation m as defined in (5.1), a simple example for geometric extrapolation is affine extrapolation: Given at least four linearly independent vertices, the affine transformation T that minimizes the sum of squared vertex distances, $\sum_{i=1}^{n_V} \|T \cdot v_i - \hat{v}_i\|^2$, can be determined by means of linear least squares fitting (see e.g. Björck (1996)). T yields an extrapolated deformation $\text{extra}_{Affine}(\text{m}) = (e : \mathbf{R}^3 \to \mathbf{R}^3, x \mapsto T \cdot x)$. Affine extrapolation is differentiable. It is invertible if and only if T is. However, in general, it is not exact on $\mathfrak{M}$.

5.1.3 Polyaffine Transformations

Given a surface mesh deformation m as defined in (5.1), and a subdivision of M into patches P_i, with $M = \cup_{i=1}^{n_P} P_i$. The extrapolation method described in this section follows the idea to (1) compute the optimal affine transformation T_i for each patch individually, and (2) merge the set of transformations into one deformation, where *local influence* of T_i at location $\mathbf{x}$ depends on the distance of $\mathbf{x}$ to patch i. Local influence of transformations is modeled by means of weight functions $w_i : \mathbf{R}^3 \to \mathbf{R}$, with $\forall x : \sum_{i=1}^{n_P} w_i(x) = 1$. A naive approach for fusion of affine transforms is to compute a linear combination:

$$e(x) = \sum_{i=1}^{n_P} w_i(x) T_i(x).$$

The resulting deformation e is differentiable if weight functions are, yet although every T_i is invertible, in general e is not, as shown e.g. by Arsigny et al. (2005).

To achieve an *invertible* fused deformation, Arsigny et al. (2005) propose to sum up *speed vector fields* $\mathcal{V}_i : \mathbf{R}^3 \to \mathbf{R}^3$ associated with each affine transformation, and define the resulting deformation as the solution of an ordinary differential equation (ODE), as detailed in the following.

Arsigny et al. (2005) propose speed vector fields $\mathcal{V}_i : \mathbf{R}^3 \to \mathbf{R}^3, x \mapsto \log(T_i) \cdot x$ where $\log(T_i)$ is the *principal matrix logarithm* of T_i. The principal matrix logarithm of a matrix A is defined via the matrix exponential:

$$\log(A) = L \Leftrightarrow \exp(L) := \sum_{k=0}^{\infty} \frac{L^k}{k!} = A.$$

As discussed e.g. by Alexa (2002), the principal matrix logarithm of a matrix A describing an invertible affine transformation $\mathbf{R}^3 \to \mathbf{R}^3$ in homogeneous coordinates is well-defined and unique if and only if A has positive determinant and real negative eigenvalues have even multiplicity. Geometrically, this is equivalent to A not describing a reflection, and not describing a rotation by π together with non-uniform scale.

$\mathcal{V}_i$ is *consistent* with T_i, i.e. integration of stationary speed vectors over time $t \in [0,1]$ starting at a position x_0 yields position $T_i \cdot x_0$: The linear, autonomous ODE

$$\dot{x}(t) = \log(T_i) \cdot x(t)$$

can be solved analytically with well-defined solutions for all t, where starting condition $x(0) = x_0$ yields solution

$$x(t) = \exp(t \cdot \log(T_i)) \cdot x_0$$

which satisfies

$$x(1) = T_i \cdot x_0 \ .$$

A weighted sum of speed vectors yields an ODE

$$\dot{x}(t) = \mathcal{V}(x(t)) := \sum_{i=1}^{N} w_i(x(t)) \mathcal{V}_i(x(t)). \tag{5.2}$$

The solution of this *polyaffine ODE* at time $t = 1$ with starting condition $x(0) = x_0$ yields the resulting deformation $e(x_0)$. As shown by Arsigny et al. (2005), e is differentiable and invertible in case all weight functions w_i are of class C^∞.

For ease of computation, Arsigny et al. (2005) propose an approximate closed-form solution of ODE (5.2): For each starting condition $x(0) = \mathbf{x} \in \mathbf{R}^3$, the solution approximating $x(1)$ is given by

$$e(\mathbf{x}) = \left(\sum_{i=1}^{n_P} w_i(\mathbf{x}) \cdot T_i^{\frac{1}{n}} \right)^n (\mathbf{x}) \ ,$$

where $n > 1$ defines a *stepsize* $1/n$. With this approximation scheme, polyaffine transformations compute a weighted sum of "n-th fractions" of each transformation, and obtain the resulting transform by concatenating the resulting sum n times. For efficient computation, $n = 2^k$ for some $k > 0$. The approximation scheme provides an *exact* solution of the ODE in case $n_p = 1$; this is because $(T^{\frac{1}{n}})^n = T$. However, in the "non-trivial" case $n_p > 1$, approximation compromises invertibility of the resulting deformation e : $\mathbf{R}^3 \to \mathbf{R}^3$. For more details, see Arsigny et al. (2005).

In summary, Polyaffine Transformations are a means of fusing patch-wise affine transformations of a surface $\mathfrak{M}$ to yield a deformation of $\mathbf{R}^3$. The resulting deformation of $\mathbf{R}^3$ is differentiable and invertible (aside from approximation errors), yet not exact w.r.t. the original deformation of $\mathfrak{M}$.

5.1.4 Mean Value Coordinates

Mean Value Coordinates (MVCs) (Floater et al., 2005) are a generalization of barycentric coordinates which allow any point $\mathbf{x} \in \mathbf{R}^3$ to be expressed as an affine combination of the vertices $\{v_i\}_{i=1}^{n_V}$ of an arbitrary closed triangular *control mesh* M (Ju et al., 2005),

$$\mathbf{x} = \sum_{i=1}^{n_V} w_i(\mathbf{x})v_i$$

where v_i are the vertices of the control mesh and $w_i(\mathbf{x}) \in \mathbf{R}$ the respective *mean value weights* at location $\mathbf{x}$, with $\forall \mathbf{x} \in \mathbf{R}^3 : \sum_i w_i(\mathbf{x}) = 1$. Note that the control mesh has to have at least four linearly independent vertices. For definition and computation of the (well-defined) mean value weight functions $w_i : \mathbf{R}^3 \to \mathbf{R}$ given a closed control mesh, see Ju et al. (2005). The control mesh does not have to be convex, nor does it have to surround $\mathbf{x}$. A deformation m of the control mesh as defined in (5.1) yields an extrapolated deformation

$$\text{extra}_{MVC}(\text{m}) := \left(\text{e} : \mathbf{R}^3 \to \mathbf{R}^3, \mathbf{x} \mapsto \hat{\mathbf{x}} := \sum_{i=1}^{n_V} w_i(\mathbf{x})\hat{v}_i \right) .$$

As shown by Ju et al. (2005), the volume deformation e extrapolates m *smoothly* inside and outside the control mesh, i.e. it is differentiable there, whereas on the control mesh itself, deformation is continuous (piecewise linear interpolation) but not differentiable. The deformation e provides an exact extrapolation of m.

Average Mean Value Weights

Consider training shapes $M^{(k)} := (V^{(k)}, E, F)$ of an anatomical structure with consistent mesh topology (cf. Sec. 3.1.2), $k = 1 \ldots n_{\mathbf{T}}$. Let $x^{(k)} \in \mathbf{R}^3 \backslash \mathfrak{M}^{(k)}$ denote a training location of an additional anatomical landmark. Mean value weights $w_i^{(k)}$ describe the location of $x^{(k)}$ relative to $M^{(k)}$. Average weights can be used to estimate a corresponding position x with respect to a deformed mesh $\hat{M}$ (e.g. obtained by SSM- plus free deformation): With v_i the vertices of the deformed mesh $\hat{M}$,

$$x := \sum_{i=1}^{n_V} \bar{w}_i v_i \quad \text{with} \quad \bar{w}_i := \frac{1}{n_{\mathbf{T}}} \sum_{k=1}^{n_{\mathbf{T}}} w_i^{(k)} . \tag{5.3}$$

Averaging mean value weights is equivalent to averaging displaced positions obtained by MVC-extrapolation of deformations of each control mesh:

$$x = \frac{1}{n_{\mathbf{T}}} \sum_{k=1}^{n_{\mathbf{T}}} \left(\sum_{i=1}^{n_V} w_i^{(k)} v_i \right) = \sum_{i=1}^{n_V} \left(\frac{1}{n_{\mathbf{T}}} \sum_{k=1}^{n_{\mathbf{T}}} w_i^{(k)} \right) v_i$$

In summary, MVCs provide for exact extrapolation which is differentiable on $\mathbf{R}^3 \backslash \mathfrak{M}$, but in general not invertible. Furthermore, averaging MVCs offers a simple and effective way of fusing extrapolations of multiple training mesh deformations onto additional training landmarks.

5.2 Atlas-based Segmentation

Atlas based segmentation (see e.g. Handels, 2009, Chapter 5.9) is an approach for segmentation of anatomical structures that does not require statistical shape modeling. It allows for implicit usage of case based prior knowledge captured by an *atlas* for segmentation of "similar" images. A medical image together with a reference segmentation, i.e. a label image, builds an *atlas*. An atlas can be seen as a "multi-object template" containing a sought structure as well as surrounding structures. Deformation of an atlas image (or *reference image*) onto a target image by means of *image registration* as described in Section 5.2.1 establishes point-to-point anatomical correspondences in 3D. Applying such a deformation to a reference segmentation of the respective atlas image yields a segmentation of target image (see Section 5.2.2 for details).

Multiple atlases yield multiple different segmentations. They can be fused into a resulting segmentation by means of *label fusion*. A straightforward method for label fusion is *majority voting*: For each voxel of the target image, for each potential label, the number of segmentations which assign this label to the respective voxel is counted. The voxel is assigned the label with highest count (and smallest ID in case a unique highest count does not exist). More sophisticated label fusion methods as well as alternative *atlas selection methods* are an active area of research (see e.g. Langerak et al. (2010)) yet lie out of the focus of this thesis.

Atlas based segmentation deforms *volumetric* "multi-object templates". Volumetric deformation models, i.e. deformation models for volumetric grids, provide a dense cross-linking of structures. In case of sparse image features and misleading imaging artifacts, this can contribute to robust segmentation of the sought structure: Not only (weak) features in the vicinity of a structure of interest, but also (more prominent) features in considerable distance can steer deformation onto the sought structure in the image. Free surface deformation models are not suitable for the task of "steering from a distance", as they do not provide for such a dense cross-linking of structures. As a consequence, they e.g. do not allow for direct control of volume changes induced onto a single structure by deformation.

As opposed to other volumetric models as e.g. FE models, atlases are comparatively easy to generate: Solely segmentations of training images are necessary; No further processing (as e.g. the generation of tetrahedral meshes) is required.

5.2.1 Image-to-image Registration

Following the categorization of Maintz and Viergever (1998), this section deals with 3D-to-3D, voxel-based image registration (see also Handels, 2009, Chapter 4). Initial (rough) registration of images is assumed to be given via a previous step in a segmentation pipeline, namely by means of mesh-based extrapolation (cf. Sec. 5.1). Hence this section is concerned with *non-parametric* registration.

The problem of finding 3d anatomical correspondences between images I and J can be formulated as a minimization problem: Find a *plausible* deformation $y :$ $\mathbf{R}^3 \to \mathbf{R}^3$ such that deformed image $I \circ y$ and image J are *similar*. Image similarity is measured by a distance function $\mathcal{D} : img \times img \to \mathbf{R}$, where $img := (\mathbf{R}^3 \to \mathbf{R})$. Minimizing image similarity without further restriction of y is an ill-posed problem (see e.g. Burger et al. (2013)). Instead, "plausibility" of y is measured with a *regularizer* function $\mathcal{S} : (\mathbf{R}^3 \to \mathbf{R}^3) \to \mathbf{R}$. The task of image-to-image registration is to find an y that minimizes $\mathcal{J}(y) := \mathcal{D}(I \circ y, J) + \mathcal{S}(y)$. Minimization of $\mathcal{J}$ w.r.t. y can be tackled by means of *numerical optimization* (Modersitzki, 2004, 2009).

The remainder of this section motivates and describes a particular regularizer and image similarity measure as applied in a segmentation pipeline in Section 10.2. It does not discuss alternative regularizers and similarity measures. To this end, see e.g. (Modersitzki, 2004, 2009).

Hyperelastic Regularizer

We aim at inter-patient registration of structures that are subject to large intra-individual variations. Initial deformation of an image by means of mesh-based extrapolation may compensate for some but not all variations: For the example of soft tissue deformations extrapolated from articulated bone deformations, bone articulation may be compensated, yet deformations that may not be compensable are e.g. muscle tension and training as well as inter-individual soft-tissue deformations not correlated to bones.

A regularizer that is able to capture large deformations and at the same time guarantees invertibility is the *hyperelastic* regularizer proposed by Burger et al. (2013). This regularizer can capture deformations which are "large" in the sense that they are not well-modeled by means of more commonly used *elastic* regularization which is based on a linear elasticity model (see Burger et al., 2013, for references). We consider this regularizer suitable for our application.

The hyperelastic regularizer has the form

$$\mathcal{S}^{hyper}(y) := \int_{\Omega} \alpha \cdot \text{length}(y) + \beta \cdot \text{volume}(y) dx \ . \tag{5.4}$$

With $y(x) = x + u(x)$,

$$\text{length}(y) := ||\nabla y - I_d||^2_{Fro} = ||\nabla u||^2_{Fro} = \sqrt{\sum_{i,j=1}^{d} (\partial_i u_j)^2}$$

This term penalizes any variation of *displacements* $u(x)$ in any direction – i.e. length and angle variations. Translations are not penalized. This is desirable for our application, as we consider local translations "anatomically plausible". As for the second term, it is defined as

$$\text{volume}(y) := f(\det \nabla y) \ , \quad \text{with} \quad f(x) := \left(\frac{(x-1)^2}{x} \right)^2 \ .$$

This term penalizes changes of volume. Note that f satisfies $f(x) = f(1/x)$. Hence both growth and shrinkage of volume are penalized. Particularly, volume(y) satisfies

$\det \nabla y \to 0 \Rightarrow \text{volume}(y) \to \infty$,
$\det \nabla y \to \infty \Rightarrow \text{volume}(y) \to \infty$, and
$\det \nabla y = 1 \Rightarrow \text{volume}(y) = 0$.

The first of these three properties is particularly important for our application: It guarantees invertibility of the deformation. In other words, folds are prevented, which is desirable w.r.t. anatomical plausibility. In summary, the regularizer $\mathcal{S}^{hyper}$, defined as a weighted sum of length- and volume term, handles deformations which are *large* in the sense that they are not well-captured by a linear elasticity model, and yields diffeomorphic transformations.

Note that Burger et al. (2013) propose an additional, third term, surface(y), which is based on the cofactor of ∇y and serves the purpose of penalizing area changes. We omit this term for our application because given explicit control over length, angle and volume changes, we do not see the need for additional, explicit control over area changes. Furthermore, omitting the surface term means omitting an additional parameter. However, this is an intuitive choice, and a thorough experimental analysis concerning the practical value of the surface term for our application remains subject to future work.

Image Similarity: Sum of Squared Intensity Differences (SSD)

For registration of images I and J acquired with the same imaging modality and exhibiting comparable intensity levels (which holds e.g. for CT), a widely used image distance measure which we also employ for our application is the sum of squared intensity differences (SSD, see e.g. Modersitzki, 2004, for references):

$$\mathcal{D}^{SSD}(I \circ y, J) := \frac{1}{2} \int_{\Omega} (I(y(x)) - J(x))^2 dx \qquad (5.5)$$

5.2.2 Application of Volume Deformations to Atlases

This section describes how volume deformations (obtained by either mesh-based extrapolation or registration or a combination of both) can be employed to deform atlas segmentations.

Given an image $I : \Omega \to \mathbf{R}$ and a deformation of its domain, $y : \Omega \to \mathbf{R}^3$. A respective "deformed image", i.e. an image I^d with $I^d(y(x)) = I(x)$, is well-defined only if y is invertible. Then, $I^d = I \circ y^{-1}$. However, what *is* well defined is a different, "reversely deformed" image, I^r, with $I^r(x) = I(y(x))$, i.e. $I^r := I \circ y$.

The goal of atlas-based segmentation is to (1) "deform an atlas image I onto an image J" such that the "deformed I" and J are similar, and (2) "transfer the deformation onto a reference segmentation of I" to yield a segmentation of J, where in general the deformation is not invertible.

There are two options for rendering I and J similar by means of a deformation of Ω: (1) Find y such that $I \circ y$ and J are similar. (2) Find z such that I and $J \circ z$ are similar. In case y is invertible, ideally, $y^{-1} = z$. In other words, ideally, "indirect" deformation of J is equivalent to "direct" deformation of I.

A reference segmentation of I which is given as a label image L_I can be deformed by means of y: Then, $L_I \circ y$ serves as a segmentation of J. In practice, to obtain a voxel segmentation of J, We compute $L_I(y(x))$ for all voxels x of J. This requires interpolation in L_I, which is performed by means of the nearest neighbor strategy (cf. Sec. 2.1.1).

A reference segmentation of I which is given as a surface mesh (V, E, F) can be deformed by means of z: Each vertex $v \in V$ is displaced to a new position $z(v)$. The resulting deformed surface mesh $(z(V), E, F)$ serves as a segmentation of J. The mesh may contain self-intersections if z is not invertible; however, even then, it can be scan-converted to yield a voxel segmentation of J

5.3   Conclusion

In this chapter, we have described state-of-the-art methods for geometric extrapolation of surface deformations onto volumes, and atlas based segmentation via 3d image registration. By no means does this chapter provide an overview of the respective fields of research. Its purpose is to describe a collection of methods that have certain desirable properties and can be of use for segmentation of anatomical structures for which SSMs are not appropriate models.

In Chapter 10 we will assemble a fully automatic segmentation pipeline that employs the methods described in this chapter. We apply it for segmentation of soft tissue structures in CT, namely muscles that are located in the vicinity of articulated bone compounds. We will present a comparative evaluation of the three extrapolation techniques described in this chapter, in combination with hyperelastic registration.

Part II

Applications to Medical Image Data

Chapter 6

Fundamentals of Quantitative Evaluation

Contents

This chapter introduces basic concepts concerning the application of methods described in Part I to clinical data, and the evaluation of respective segmentations. The goal of quantitative evaluation of segmentation accuracy is to make statements as to how accurate a method is, and whether method A is more accurate than method B. As for the quantification of segmentation accuracy, Section 6.1 introduces a set of measures that quantify the dissimilarity of one segmentation to another. These measures quantify differences between segmentations. They serve for assessing errors of automatic segmentations w.r.t. manual reference segmentations as well as differences between manual segmentations. Quantitative evaluation of segmentation accuracy is usually performed on a pool of *test data*. Section 6.2 introduces common ways of presenting sets of error measurements assessed on a pool of test data. Section 6.3 discusses the comparability of different segmentation methods which target the same application. Section 6.4 discusses the issue of *over-fitting*. Section 6.5 discusses application-specific parametrizations of methods as performed in the following chapters.

6.1   Measures of Segmentation Accuracy

In the following we will detail a set of "standard-" dissimilarity measures (cf. e.g. van Ginneken et al. (2007) and Handels (2009), Chapter 5.10), namely the *Dice coefficient*, the *Jaccard coefficient*, and the *relative volume difference* (which are *volumetric* measures), as well as *mean, root mean square* and *Hausdorff* asymmetric and symmetric *surface distances*.

Given two voxel segmentations to be compared, let P, Q denote the respective subsets of $\mathbf{R}^3$ labeled as "foreground". The volume of P can be computed easily by counting labeled voxels and multiplying by voxel volume. We denote it as $vol(P)$. The Dice coefficient is computed as

$$\text{Dice}(P, Q) := \frac{2 \cdot vol(P \cap Q)}{vol(P) + vol(Q)} .$$

The Jaccard coefficient is computed as

$$\text{Jaccard}(P, Q) := \frac{vol(P \cap Q)}{vol(P) \cup vol(Q)} .$$

Both Dice and Jaccard coefficient lie within a range $[0, 1]$, where a value 1 indicates perfect match (equality) of segmentations P and Q. The coefficients can be transformed into respective *overlap errors* by subtraction from 1. Then, a value of 0 indicates perfect match. If the preferred unit is percent, the respective coefficient or overlap error can be multiplied by 100. In the following, we denote Dice and Jaccard overlap error as DOE and JOE, respectively.

The relative volume difference is computed as

$$\text{RVD}(P, Q) := \frac{vol(P) - vol(Q)}{vol(Q)} .$$

This is an *asymmetric* measure, where Q is commonly a reference segmentation. It lies within a range $[-1, \infty)$. A value smaller than zero indicates that the volume of P is smaller than the volume of Q. A value of zero indicates equal volume, but does not contain any information about the overlap of segmentations.

Surface distance measures are based on Euclidean distances between boundaries of P and Q. We represent these boundaries as triangle meshes (cf. Sec. 2.1.4), denoted as M_P and M_Q in the following, with vertex lists V_P and V_Q, respectively. This allows for a discrete formulation of surface distance measures. As for *asymmetric* distance measures *from M_P to M_Q*, the *mean surface distance* is defined as

$$\text{md}(M_P, M_Q) := \frac{1}{|V_P|} \sum_{v \in V_P} \min_{x \in \mathfrak{M}_Q} ||v - x|| ,$$

the *root mean square surface distance* is defined as

$$\mathrm{rmsd}(M_P, M_Q) := \sqrt{\frac{1}{|V_P|} \sum_{v \in V_P} \min_{x \in \mathfrak{M}_Q} ||v - x||^2} \; ,$$

and the *Hausdorff distance* or *maximum surface distance* as

$$\mathrm{hd}(M_P, M_Q) := \max_{v \in V_P} \left(\min_{x \in \mathfrak{M}_Q} ||v - x|| \right) \; .$$

Each asymmetric surface distance measure comes with a symmetric version, where

$$\mathrm{MD}(M_P, M_Q) := \tfrac{1}{2} \left(\mathrm{md}(M_P, M_Q) + \mathrm{md}(M_P, M_Q) \right) \; ,$$

$$\mathrm{RMSD}(M_P, M_Q) := \sqrt{\tfrac{1}{2} \left(\mathrm{rmsd}(M_P, M_Q)^2 + \mathrm{rmsd}(M_P, M_Q)^2 \right)} \; \text{ and}$$

$$\mathrm{HD}(M_P, M_Q) := \max \left(\mathrm{hd}(M_P, M_Q), \mathrm{hd}(M_P, M_Q) \right) \; .$$

Each error measure reveals particular characteristics of a segmentation. For evaluation of segmentation accuracy, usually, not one single, but a set of these scalar error measures is assessed. For a thorough discussion of the particularities of individual error measures, see Niessen et al. (2000).

To measure the accuracy of a segmentation *relative to the inter-observer variability* inherent to a specific segmentation task, van Ginneken et al. (2007) propose a *scoring system*: Jaccard overlap error, relative volume difference, and all symmetric surface distance measures defined above are assessed w.r.t. a ground truth segmentation. The values are transformed linearly to a range $[0, 100]$ relative to the respective values assessed for second expert segmentations: An exact hit of the ground truth "scores" 100, errors equal to the average distance of second observers to the ground truth score 75, and so forth (double errors score 50, triple errors score 25, ...). There are no negative scores – errors of more than four times the average second observer distance are considered *failures* and score 0. The *overall score* is the average of the five measure-wise scores.

Neither of the above measures nor their combination takes into account the location of dissimilarities. Information about locations of errors is captured by individual or averaged *surface distance maps*. An example is shown in Figure 6.1.

Note, *voxel segmentations* of images (i.e. voxel labelings, cf. Sec. 2.1.2) can only *approximate* the "true" spatial extension of anatomical structures due to their digital nature: A voxel gets one label only and hence partial volume effects occur (cf. Sec. 1.2.1). In this sense, a voxel segmentation is inherently inaccurate with surface distances up to one half voxel size. Volumetric error measures are directly computed from voxel segmentations. In contrast, surface distance measures are

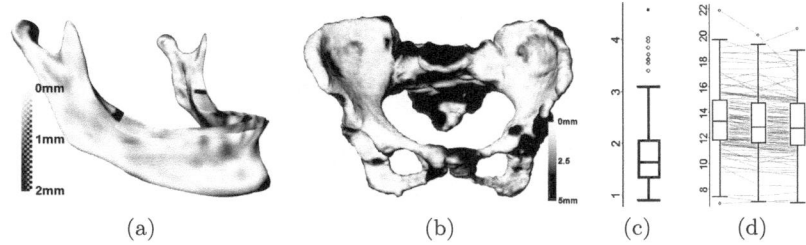

(a)   (b)   (c)   (d)

Figure 6.1 Surface distance map: Distance for each point on the surface of an automatic segmentation to the respective ground truth surface, shown for (a) an exemplary mandible and (b) an exemplary pelvis segmentation result. (c) Exemplary box plot. Y-Axis: Unit of an error measure. (d) Exemplary side-by-side box plots for different methods with under-laid parallel coordinate plots.

computed from triangle surface meshes. Reference surface meshes are directly generated from voxel segmentations (cf. Sec. 2.1.4). On the other hand, surface meshes yielded by automatic segmentation methods are generated by mesh deformation. Hence they do not directly stem from voxel segmentations. While such meshes potentially allow for image segmentation with *sub-voxel accuracy*, as mentioned in Section 2.1.2 a respective discussion lies out of the focus of this thesis.

6.2   Presentation of Results

In the following chapters we assess the accuracy of segmentations of images stemming from pools of test data. For this purpose we employ multiple error measures. Multiple error measures assessed for a number of test data can be presented as tables listing each measure for each test case.

Box plots (Chambers, 1983) allow for a more concise presentation by means of *error statistics*. A box plot gives an overview on one error measure assessed for a number of test cases. It shows statistics of the respective set of measurements, namely median and quartiles, outlier-free range of measurements, and outliers. Following Chambers (1983), we illustrate the inter-quartile range as a *box*, i.e. the outline of a rectangle, the outlier-free data range as *whiskers* (i.e. vertical lines protruding from the box on both sides and ending in horizontal lines), the median as a horizontal line through the box, and the average as a small diamond. *Outliers* are depicted as small circles, and *extreme outliers* as small squares. (See Chambers (1983) for respective definitions.) See Figure 6.1c for an exemplary box plot.

Alternatively, to boil down a set of measurements to just a couple of quantities, we report the average and standard deviation of the set of measurements.

6.3   Comparison of Methods

Direct comparison of segmentation algorithms published by different authors is difficult (see e.g. Heimann et al. (2010)). This is because usually different authors evaluate their methods on different data. The data most often stems from different sources, exhibits different resolution, different states of pathology, and so on. Comparing accuracy as achieved by different methods in terms of the same error measure, yet on different data, allows for no more than a weak hint at differences between methods. Even this indirect comparison is not always possible, as different authors often employ different accuracy measures for evaluation.

Providing a common pool of *benchmark* test data as well as establishing standard measures for evaluation is indispensable for direct comparison of methods for automatic image segmentation. An initiative that pushes in this direction since 2007 are the *MICCAI Grand Challenge* workshops (see e.g. van Ginneken et al. (2007); Heimann et al. (2010) and www.grand-challenge.org). Thanks to this initiative, benchmark data is available e.g. of liver CT and knee MRI. On this data, our segmentation methods are directly comparable to related work in terms of a *score* that measures accuracy (cf. Sec. 6.1).

Direct Comparison of Methods

Given a common pool of test data as well as error measures, *hypothesis tests* allow for a *statistical* comparison of methods. Thus statements can be made concerning the significance of differences between methods. A hypothesis test concerning differences between methods computes the probability that the *null hypothesis* holds, i.e. that methods are *not* different, despite the observed data. This probability is called *p-value*. If a p-value lies below a pre-defined *significance level*, commonly five or one percent, it indicates the respective null hypothesis is highly unlikely.

Given error measurements for two different methods on a common pool of test data and w.r.t. the same error measure, each test case yields a pair formed by both methods' respective error measurements. *Paired tests* analyze sets of pairwise differences. (This is opposed to unpaired tests, where sets of method-individual measurements are assumed to be independent.) *Wilcoxon's signed-rank test* (see e.g. Hollander and Wolfe (1999)) is a *non-parametric* paired test, i.e. it does not make assumptions on the distributions of error measurements. It can be used instead of paired *t-tests* in case the data cannot be assumed to be normally distributed. The normality hypothesis can be tested with the *Shapiro-Wilk test* (see e.g. Sen and Srivastava (1990)). Usually it does not hold on our test data, and hence we employ Wilcoxon's signed-rank test to assess the significance of differences between methods.

Side-to-side box plots allow for a visual comparison of error measurements for different methods. In case of a common pool of test data and common error measure, *parallel coordinate plots* directly visualize pairs of errors. Parallel coordinate plots draw lines between errors measured for different methods on corresponding test cases. Figure 6.1d shows exemplary side-by-side box plots under-laid with parallel coordinate plots.

6.4 Generalization to New Image Data

As opposed to test data, *Training data* is a set of exemplary data from which a method learns prior knowledge, be it a statistical shape model, an intensity model, an algorithm for individual parametrization of a heuristic intensity model, or any other knowledge. If training and test data are not separated, there is a risk of *overfitting* (see e.g. Sammut and Webb (2011)): A method may perform accurately on the available data, but considerably worse on previously unseen data. In this regard the term *generalization performance* refers to the performance of a method (w.r.t. accuracy) on data not used for model training.

The *leave-one-out cross-validation* strategy (see e.g. Sammut and Webb (2011)) allows for training from almost all available data and at the same time clearly separates training and testing data: With D denoting the set of available data, leave-one-out cross-validation proceeds as described in Algorithm 3. Note that this

---

Algorithm 3 Leave-one-out cross-validation

  for all $d \in D$ **do**
    Train method, i.e. learn prior knowledge, from $D \setminus \{d\}$.
    Apply trained method to d and evaluate result.
  end for

---

strategy is a special case of *cross-validation*, namely cross-validation with number of folds equal to the number of training data.

Quantitative evaluation allows for statements of accuracy and significance of differences between methods *on test data*. Concerning new data that appears in clinical practice, there is never any *guarantee* as to how well a method will work on this data. However, separating training and test data at least allows for a glimpse on the generalization performance of a method.

6.5 Parameter Settings

Methods as described in Part I have *parameters* that need to be set for each application. Parameters are e.g. length and sampling distance of unidirectional displacements (cf. Sec. 3.3.1), the number of modes considered in an SSM (cf. Sec. 3.1.1), intensity thresholds (cf. Sec. 3.2.1), the number of Gaussians in a mixture model fitted to an intensity histogram to determine thresholds (cf. Sec. 3.2.2), etc.

In this thesis, if not stated otherwise, we set parameters to application-specific values that we determine *heuristically*, i.e. from experiments with respective image data. In other words, we *manually tune* parameters per application with the goal of achieving good segmentation accuracy. However, once set, parameters are fixed per application, which makes our segmentation pipelines fully automatic. We state the application-specific parameter values in the respective Application sections.

We usually tune parameters from experiments with all data available for an application. As parameter tuning is a way of training prior knowledge, this approach violates the separation of training and test data. However, for datasets containing about 50 or more images, we consider this a slight violation: We hypothesize that manual tuning of about a handful of parameters will not result in a pronounced over-fitting of such data. This hypothesis is supported by an evaluation as performed in an on-site segmentation contest, where the respective test data was not available off-site (cf. Sec. 8.2): Accuracy on this data is comparable to accuracy achieved on previously available data. However, another on-site segmentation contest reveals over-fitting of particular data containing pathologies of which only a small number (< 5) was available for training and testing (cf. Sec. 7.1).

Future work will investigate methods for tuning parameters automatically while maintaining a clear separation of training and test data. To this end, a framework called *Tuner* (Torsney-Weir et al., 2011) has been shown to yield promising results.

Chapter 7

Single-object Segmentation of Anatomical Structures

Contents

In this chapter, we propose fully automatic segmentation pipelines for three segmentation problems that appear in clinical practice. Pipelines are assembled from methods described in Chapter 3 (Sections 3.1 - 3.4).

In Section 7.1, we present a fully automatic segmentation method for the liver in contrast-enhanced CT, as published in Kainmueller et al. (2007) (see also Heimann et al. (2009)). It is based on a combination of SSM- and shape-constrained free deformations. The liver exhibits large inter-individual shape variations: An SSM can only provide relatively rough approximations of individual shapes as compared to more rigid, i.e. less varying structures like bones (Heimann and Meinzer, 2009). Consequently, for accurate liver segmentations, SSM deformations need to be followed by free deformations that allow for high flexibility. The FreeBand approach described in Section 3.4.1 provides this kind of flexibility, while remaining robust w.r.t. *leakage* (cf. Sec. 1.2.1) via its narrow-band shape constraint. Mesh adaptation to image data is performed according to a simple heuristic appearance model that captures typical intensity patterns around the liver boundary and considers the potential presence of tumors in the liver. With this method, we won the first prize at the MICCAI *3D Segmentation in the Clinic: A Grand Challenge* on-site liver segmentation contest in 2007. As at June 2014, our method is the most accurate among all 21 automatic methods competing on a benchmark pool of liver CTs in terms of a *score* (cf. Sec. 6.1) that measures segmentation accuracy (cf. www.sliver07.org).

In Section 7.2, we present an algorithm for automatic segmentation of the pelvic bones from CT as published in Seim et al. (2008). As with the liver, the method is based on SSM- and subsequent free deformations. Intensity characteristics of pelvic CT call for free deformations that are robust w.r.t outliers and misleading image features (cf. Figure 3.2). The *GraphCuts* method as presented in Section 3.4.2 provides this kind of robustness. We present a step-wise evaluation on 50 CTs of the pelvic region. Evaluation reveals high accuracy compared to related work, yet results are not directly comparable due to a lack of benchmark data.

Section 7.3 deals with the exact localization of the mandibular bone and nerves in cone beam computed tomography (CBCT). CBCT is increasingly utilized in maxillofacial or dental imaging (Schramm et al., 2005). Compared to conventional CT, however, soft tissue discrimination is worse due to a reduced dose. Thus, small structures like the alveolar nerve channels are hardly recognizable within the image data. We show that accurate fully automatic segmentation of mandibular bones as well as delineation of alveolar nerve channels is nonetheless possible. To this end, as published in Kainmueller et al. (2009b), we propose a method that is based on a *compound SSM* of bone and nerves, where we extrapolate nerves from bone as described in Section 3.1.5. We achieve image-based refinement of SSM-extrapolated nerve delineations with an application-specific approach that involves Dijkstra's Algorithm (Dijkstra, 1959). We evaluate our method on 106 clinical datasets. To the best of our knowledge, our work presents the first quantitatively evaluated fully automatic 3d method for segmentation of mandibular bone and nerves in CBCT.

In this chapter we do not focus on the run-time of segmentation pipelines. However, for sake of completeness note that average run-times range between four minutes (pelvis) and 15 minutes (liver) per test case on a 3.2 GHz Core.

7.1 Segmentation of the Liver in Contrast-enhanced CT

In contrast-enhanced CT, intensity values of liver tissue depend on the uptake of contrast agent, which in turn depends on the time-point at which the image is taken. Consequently, liver intensities vary among individual images. In the pool of image data we could access for our work, we observed average intensities ranging from about 50 to about 250 HU. Intensities are often similar to those of surrounding anatomical structures like the stomach, heart, pancreas, kidney and muscles (see also e.g. Heimann et al. (2009); Mharib et al. (2012)), some of which also take up contrast agent. Consequently, low-level segmentation approaches which are based solely on local intensity or intensity gradient features are usually not sufficient to differentiate between liver tissue and adjacent anatomical structures.

Prior knowledge about the typical shape of a liver can be incorporated into the segmentation process to constrain it where image information is not reliable. The shape may be constrained by a single template (Montagnat and Delingette, 1997), an SSM (Cootes et al., 1994) or more flexible deformable models. Combinations of these approaches have also been presented (Weese et al., 2001; Heimann et al., 2007). Building upon the work of Lamecker et al. (2002, 2004b), our method adopts a combination of SSM- and shape-constrained deformable models as described in Sections 3.1 and 3.4.1, respectively. Section 7.1.1 describes our liver SSM.

Just as important and challenging as shape modeling itself is the process that drives the shape model to match the image data to be segmented. General intensity features (Montagnat and Delingette, 1997; Weese et al., 2001) or a statistical model of the intensity distribution (Heimann et al., 2006) have been used. In Section 7.1.3 we propose a heuristic model of typical image appearance around the liver boundary. This appearance model is an application-specific modification of the generic model presented in Section 3.2. It considers the potential presence of tumors inside the liver. An algorithm for computing an appearance match displacement field is derived. This algorithm implicitly defines a binary appearance cost function which chooses no more than one displacement per vertex to match within a respective set of candidate displacements, i.e. $\forall v \in V : \exists_1 l \in L(v) : \phi(v, l) = 1$ (cf. Sec. 3.2.1). This method extends the work of Lamecker et al. (2002, 2004b).

In order to achieve fully automatic segmentation, an initial position of the liver SSM in the image data as well as parameters of the appearance model need to be estimated automatically. Our solutions for these problems are presented in Sections 7.1.2 and 7.1.3. Section 7.1.4 describes our fully automatic liver segmentation pipeline. An evaluation on a pool of 101 liver CTs is presented in Section 7.1.5, together with a discussion of results.

7.1.1 Statistical Shape Model of the Liver

The SSM of the liver we employ for segmentation was generated from 102 training shapes, with the methods described in Sections 3.1.1 and 3.1.2. As training data we used manually segmented PACS data (71 CTs and 11 MRIs) and additional 20 training datasets as provided by the MICCAI workshop *3D Segmentation in the Clinic: A Grand Challenge* (see van Ginneken et al. (2007) and www.sliver07.org). All training datasets were manually labeled by experts in liver anatomy. Point-to-point correspondences between training liver surfaces were established via consistent subdivision of all surfaces into five patches. The resulting SSM has about 7000 vertices.

7.1.2 Application-specific Initialization

We do not employ the Generalized Hough Transform as presented in Section 3.1.4 for initialization of the liver SSM because the average liver shape may capture "too little" of an individual liver shape to yield satisfactory results. However, an extension of the GHT as proposed by Ruppertshofen et al. (2011) may provide a generic alternative to our application-specific initialization method as described in the following. A respective evaluation is subject to future work.

The idea of our application specific method for position initialization is to robustly detect the lower rim of the right lobe of the lung and to position the liver model below it. We assume that patient orientation in a CT scanner is encoded in the DICOM header (e.g. "Feet First Supine"). First, all connected components with intensity values less than -600 HU are determined in the image. The largest two components (left and right lobe) which are adjacent to the upper border of the image volume are selected. Next, the lobe component on the right-hand side is projected in patient axis direction from feet to head and the center and orientation of the resulting lung area is determined. The liver model is translated and oriented according to the back-projected center point and orientation of the lower rim of the right lobe.

7.1.3 Heuristic Appearance Model for Displacement Computation

We assume liver tissue to appear homogeneous in contrast-enhanced CT, and to exhibit a distinct gradient to some surrounding structures. To filter out noise while maintaining sharp gradients, we apply nonlinear isotropic diffusion filtering to the image (Weickert et al., 1998). Figure 7.1 shows a slice of contrast-enhanced liver CT before and after filtering.

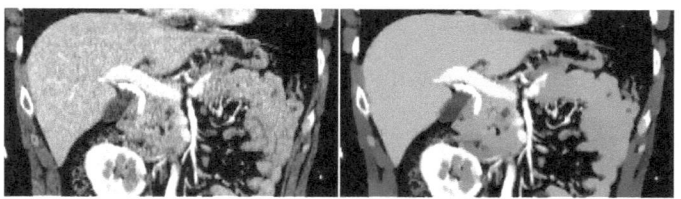

Figure 7.1 Slice of liver CT before (left) and after (right) filtering with a non-linear isotropic diffusion filter.

Figure 7.2 shows typical intensity profiles for different regions on a liver surface in filtered image data. We employ a simple heuristic *appearance model* to describe the intensity distribution around the liver boundary in diffusion filtered images: Intensities inside the liver are assumed to lie within a window $W^{liver} = [\bar{w}^{liver} - r^{liver}, \bar{w}^{liver} + r^{liver}]$, where $\bar{w}^{liver}$ denotes average liver intensity and r^{liver} some tolerance "radius". Analogously, intensities of tumors potentially contained in the liver are assumed to lie within a window $W^{tumor} = [\bar{w}^{tumor} - r^{tumor}, \bar{w}^{tumor} + r^{tumor}]$. Furthermore, intensity thresholds t_{min} and $t_{max} := \bar{w}^{liver} \pm 3r^{liver}$ serve as boundaries below resp. above which intensities are assumed to definitely indicate non-liver and non-tumor tissue. The liver boundary is assumed to be characterized by image derivatives in surface normal direction with an absolute value above a threshold g. On an intensity profile with n sample points, a sample point x is only considered to lie inside liver or tumor tissue if it belongs to a consecutive set of at least $c_{min} := n/5$ candidate locations along the profile that lie within W^{liver} or W^{tumor}, respectively.

Estimating Appearance Parameters

The intensity parameters of the appearance model are estimated based on analyses of two histograms, H_1 and H_2, of the filtered image data. The basic idea is to discriminate the major "liver peak" from minor yet distinct other peaks which indicate the presence of tumor tissue. H_1 is the histogram of voxel intensities inside some liver surface mesh. Since Histogram H_1 possibly excludes important regions of the liver which are not covered by an intermediate segmentation, we also analyze a second Histogram, H_2, which considers a slightly *enlarged* volume by inflating the current liver surface by 10 mm. The histograms are evaluated only on voxels with an intensity in the range of $[0, 300]$ HU, which we assume to cover both liver and tumor tissue. A weighted sum of 10 Gaussians $gauss_i$ $(i = 1, \dots, 10)$ is fitted to each of the two histograms using the Expectation Maximization (EM) algorithm on a Gaussian mixture model (cf. Sec. 3.2.2). As a result, we obtain weights w_i, means μ_i, standard deviations σ_i and *peak heights* $h_i := w_i/\sigma_i$ for each Gaussian.

Algorithm 4 Tumor intensity range

1: Of the 10 Gaussians fitted to H_1, identify the one with highest peak: $p := \text{argmax}_i \{h_i\}$. It is assumed to model the intensity distribution inside the liver.

2: Identify *nearby* Gaussians whose means are at least 15 and at most 20 HU away and whose peaks are higher than 2% of h_p: $N := \{j : |\mu_p - \mu_j| < \max(15, \min(20, 3\sigma_p)), h_j > 0.02 \cdot h_p\}$

3: Set a temporary *lower liver boundary*: $l_1 := \min_{j \in N}\{\mu_j - \sigma_j \cdot \max(1, \min(3, 50 \cdot h_j^2/h_p^2))\}$. Set an *upper liver boundary* u_1 accordingly.

4: Compute *total liver peak height* $h := \sum_{j \in N} h_j$. Note that h is not necessarily a peak height of our Gaussian mixture.

5: Now identify *potential tumor peaks* with means darker than the lower liver boundary and peak heights exceeding 5% of h: $T := \{j \mid \mu_j < l_1, h_j > 0.05 \cdot h\}$.

6: If $|T| = 0$, assume that there is no big tumor. Return.

7: Otherwise identify $\sigma_{\min} := \min_{j \in T}\{\sigma_j\}$. Set $T^* := T \setminus \{j \in T \mid \sigma_j > 2\sigma_{\min}\}$ and $t := \text{arg max}_{j \in T^*}\{\mu_j\}$. Gaussian t is assumed to model the intensity distribution inside tumors.

8: Set $\bar{w}^{tumor} := \mu_t$ and $r^{tumor} := \min(20, 3\sigma_t)$.

First, the tumor intensity range W^{tumor} is estimated as described in Algorithm 4. Tumor intensity estimation is based on histogram H_1, since an enlarged volume as considered by H_2 may capture soft tissues surrounding the liver whose intensities may be mistaken for tumor tissue, like e.g. muscle or fat. Algorithm 4 also computes a first estimate of the liver intensity range.

In a second step we analyze the Gaussians fitted to H_2 to see whether the liver intensity range needs to be enlarged, as described in Algorithm 5.

Algorithm 5 Liver intensity range

1: Derive liver intensity boundaries l_2 and u_2 from H_2 just like l_1 and u_1 from H_1.

2: Identify a tumor Gaussian $\tilde{t}$ for H_2 just as t for H_1. Accept $\tilde{t}$ only if no t was found, and if $\mu_{\tilde{t}} > 0.5\,(l_2 + u_2) - 30$. Then set $\bar{w}^{tumor} := \mu_{\tilde{t}}$ and $r^{tumor} := \min(20, 3\sigma_{\tilde{t}})$. This reduces the risk of including muscle tissue in the liver range.

3: Set $u := \max(u_1, u_2)$. If $l_1 < l_2$ and not $\bar{w}^{tumor} > l_1$, set $l := l_1$. Otherwise set $l := l_2$.

4: Fix $\bar{w}^{liver} := 0.5\,(l + u)$ and $r^{liver} := 0.5\,(u - l)$, and $g := 0.5\,r^{liver}$.

In summary, Algorithms 4 and 5 estimate image-individual liver intensity parameters $\bar{w}^{liver}$, r^{liver} and g, decide whether there is a large tumor inside the liver, and if yes, estimate tumor intensity parameters $\bar{w}^{tumor}$ and r^{tumor}. Algorithms are *heuristic* (cf. Sec. 6.5) in that we developed and parametrized them based on observations of image data together with initialized and deformed liver meshes.

Computing the Displacement Field

For each vertex v of the surface mesh we analyze a 1D intensity profile over a length $2r$ along the surface normal n_v (cf. Sec. 3.3.3). Fig. 7.2 shows some typical intensity profiles in different anatomical regions. For a vertex v_i, the domain of a profile is a set of n equidistant candidate locations, L_i, as described in Section 3.3. The goal of intensity profile analysis is to assign a displacement $l_a \in L_i, 1 < a \leq n$, and a confidence weight $w(v_i)$ to each vertex v_i of the surface. We propose to compute l_a and w from an intensity profile at a vertex as described in Algorithm 6. In summary, Algorithm 6 decides whether a profile runs through liver or tumor tissue, and if yes determines the liver boundary as a point of inflection on the profile that exhibits some reasonable intensity. Note that this algorithm implicitly defines a binary cost function $\phi : V \times S \rightarrow \{0, 1\}$ with zero cost for exactly one displacement (namely l_a), i.e. $\forall v_i \in V : \exists_1 l \in L_i : \phi(v_i, l) = 0$ (cf. Sec. 3.2.1). The weight $w(v_i)$ encodes how "sure" we are about this displacement.

Algorithm 6 Liver intensity profile analysis

1: Initialize $a := (n + 1)/2$ and $w := 0$.

2: Determine largest i with $n \geq i \geq c_{\min}$ for which $I(v + l_{i-k}) \in W^{liver}$ for all k with $0 \leq k < c_{\min}$. If there is such an i, set $a := i$ and $w := 1$, and define $v + l_a$ to be inside liver.

3: If $v + l_a$ not inside liver $\rightarrow$ If tumor present, determine largest i with $n \geq i \geq c_{\min}$ for which $I(v + l_{i-k}) \in W^{tumor}$ for all k with $0 \leq k < c_{\min}$. If such an i exists, set $a := i$ and $w := 0.75$, and define $v + l_a$ to be inside tumor.

4: If $v + l_a$ neither inside liver nor inside tumor $\rightarrow$ Determine smallest $i \leq n$ with $I(v + l_j) < t_{\min}$ for all j with $i \leq j \leq n$. If such an i exists and $i < a$, set $a := i$, $w := 0.75$ and return.

5: If $w = 0 \rightarrow$ Return.

6: Now $v + l_a$ is either inside liver or inside tumor. The remaining steps are equal in both cases. $\rightarrow$ Find first point $i > a$ with either $|\nabla_{n_v} I(v + l_i)| > g$ or $I(v + l_i) \notin W^{liver/tumor}$. If $I(v + l_i) \notin W^{liver/tumor}$, set $a := i - 1$, else $a := i$.

7: If $|\nabla_{n_v} I(v + l_a)| > g \rightarrow$ Find point of inflection $v + l_i$ with smallest $i > a$, then find largest $k < i$ such that $I(v + l_k) > t_{\min}$. Set $a := k$.

8: Find largest $i \leq n$ with $I(v + l_i) < t_{\max}$. If $i < a$, set $a := i$. This prevents from moving too far into bright regions such as kidney or heart.

7.1.4 Segmentation Pipeline

The pipeline for fully automatic segmentation of the liver is described in Algorithm 7. Its core components are position initialization (**Init**, see Sec. 7.1.2), parameter estimation (**Estimate**, see Sec. 7.1.3), SSM deformation (**SSM**, see Sec. 3.1.3),

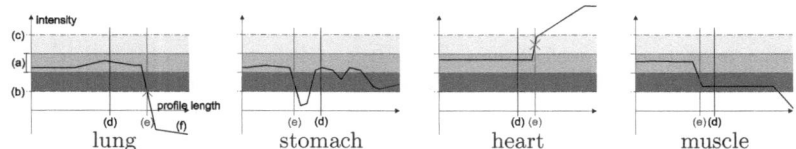

lung   stomach   heart   muscle

Figure 7.2 Exemplary intensity profiles. (a) W^{liver}, (b) $t_{\min}$, (c) $t_{\max}$, (d) current vertex position v, (e) new position $v + l_a$ as computed by Algorithm 6, (f) profile plot.

and shape-constrained free deformation (**FreeBand**, see Sec. 3.4.1). For each step specific parameter settings and conditions are specified in the algorithm.

In summary, Algorithm 7 performs a sequence of mesh deformation steps with increasing flexibility in terms of degrees of freedom of the deformation model. We assume the mesh gets closer to the target surface over time. Hence we (1) decrease profile length accordingly, and (2) repeatedly re-estimate intensity parameters from potentially more accurate liver volumes. In Step 8 of the algorithm, the maximum number of used shape weights depends on the slice thickness of the original CT data. This setting results from the observation that using less shape weights yields better segmentations on data with coarse resolution. We hypothesize that the lack of information in coarse resolution data is better coped for by restricting the statistical shape model to "more probable" shapes. As for Step 12, the purpose of re-meshing is to provide FreeBand with an intersection-free initial surface mesh. In consequence, FreeBand is guaranteed to yield an intersection-free mesh as well. Step 15 Fills 2D holes in xy-slices of the resulting label field which makes the resulting segmentation conform to the gold standard segmentation protocol for the vena cava.

7.1.5 Results and Discussion

Comparisons of automatic to manual segmentations were performed on 10 benchmark liver CTs provided by the MICCAI workshop *3D Segmentation in the Clinic: A Grand Challenge* (see van Ginneken et al. (2007) and www.sliver07.org). Tab. 7.1 lists quantitative results in terms of Jaccard overlap error (JOE), relative volume difference (RVD), symmetric mean, root mean square and Hausdorff surface distance (MD, RMSD, HD), and *score* as introduced in Section 6.1 (see also van Ginneken et al. (2007)). Our method achieves an average overall score of 77.3 and ranks first among all competing automatic methods (see www.sliver07.org).

We performed an additional evaluation on 91 liver CTs, of which 71 were provided via a PACS, and 20 were provided as training data for the MICCAI workshop *3D Segmentation in the Clinic: A Grand Challenge* (see www.sliver07.org).

Algorithm 7 Liver segmentation. Parameters stay the same unless noted otherwise.

1: **Preproc:** Nonlinear diffusion filtering of I
2: **Init** position T^0
3: **Estimate** W^{liver}, W^{tumor}, g
4: **SSM:** Adapt position only (rigid + isotropic scaling $\in [0.5, 1.5]$), profile length $2r := 50\,\text{mm}$, number of sample points $n := 51$, $t_{\min} := 0$, convergence criterion $\epsilon := 0.1\,\text{mm}$.
5: **Estimate:** Recompute W^{liver}, W^{tumor}, g
6: **SSM:** Adapt position and 5 shape parameters. If no tumor, $2r := 60$ mm and $n := 61$.
7: **Estimate:** Recompute W^{liver}, g
8: **SSM:** Adapt position and m shape parameters. If no tumor detected, $m := min(50, max(20, -7.5z + 57.5))$, with z the slice thickness of the original CT in mm, and $t_{\min} := \bar{w}^{liver} - 3r^{liver}$. Otherwise $m := \min(30, \max(20, -7.5z + 57.5))$ and $t_{\min} := \bar{w}^{tumor} - r^{tumor}$. $\epsilon := 0.05\,\text{mm}$
9: **Estimate:** Recompute W^{liver}, g
10: **SSM** with $2r := 40$ mm and $n := 41$
11: **SSM** with $2r := 20$ mm, $n := 41$, only performed if no tumor
12: **Re-mesh:** Scan convert deformed SSM into binary labeling, fill interior holes of foreground, generate a new surface mesh as input for the following steps.
13: **FreeBand** with $2r := 30$ mm, $n := 61$, $\epsilon := 0.04$ mm, narrow band radius $:= 10$ mm, maximum number of iterations $:= 30$, only performed if no tumor.
14: **FreeBand** with $2r := 10$ mm, $n := 51$
15: **FillSlices:** Fill 2D holes in xy-slices of scan converted result from Step 14

Test case	JOE [%]	Scr	RVD [%]	Scr	MD [mm]	Scr	RMSD [mm]	Scr	HD [mm]	Scr	Total Scr
1	6.03	76.4	-2.08	88.9	0.93	76.8	2.10	70.8	20.25	73.4	77.3
2	9.79	61.8	-8.92	52.6	1.31	67.2	1.89	73.8	17.04	77.6	66.6
3	4.30	83.2	-0.65	96.5	0.83	79.2	1.62	77.5	18.70	75.4	82.4
4	5.89	77.0	0.06	99.7	0.92	77.0	1.81	74.9	16.38	78.4	81.4
5	5.70	77.7	-1.88	90.0	0.93	76.8	1.67	76.8	18.85	75.2	79.3
6	7.71	69.9	-5.57	70.4	1.48	63.0	3.83	46.8	41.36	45.6	59.1
7	3.24	87.3	-0.77	95.9	0.44	89.0	1.03	85.7	11.36	85.1	88.6
8	5.14	79.9	-2.66	85.9	0.81	79.8	1.63	77.4	12.25	83.9	81.4
9	4.10	84.0	-0.57	97.0	0.47	88.2	0.93	87.1	14.65	80.7	87.4
10	8.96	65.0	-5.57	70.4	1.34	66.5	2.22	69.2	16.08	78.8	70.0
Avg	6.09	76.2	-2.86	84.7	0.95	76.3	1.87	74.0	18.69	75.4	77.3

Table 7.1 Error metrics and scores (Scr) for automated segmentations of ten benchmark CTs. Last row: Average errors and scores.

We employed the liver SSM in a leave-one-out manner. On the total of 101 liver CTs, comparing automatic segmentations to manual reference segmentations yields the following average error measures: JOE 7.3%, RVD 0.9%, MD 1.3 mm, RMSD 2.4 mm, HD 20.7 mm. The maximum error measures occurring among all 101 cases are the following: JOE 15.6%, RVD -8.5 and +10.7%, MD 2.6 mm, RMSD 5.5 mm, HD 44.1 mm. This corresponds to an average score of 72.7. The 101 CTs exhibit slice thicknesses between 1 and 5 mm. When considering only CTs with a slice thickness below 3 mm, the average mean surface distance (MD) reduces to 0.9 mm.

In an on-site segmentation contest held at MICCAI 2007 workshop *3D Segmentation in the Clinic: A Grand Challenge*, we achieved an average score of 68 on another set of 10 CTs that were not made available before (see Heimann et al. (2009)). Results on this data are not prone to a potential over-fitting by means of manual parameter tuning (cf. Sec. 6.5). As for most other competing methods our average score on this data is considerably inferior to the one achieved on the 10 test datasets (which is 77.3, see Table 7.1). Heimann et al. (2009) report respective test- and on-site scores of all competing methods. In our case the drop in score is due to one particular on-site CT with a large tumor where tumor detection failed. This hints at an over-fitting of our tumor detection method (see Algorithm 4) to the few data (less than five CTs) with large tumors that were available for training and testing before the on-site contest.

In terms of score, a value above 75 indicates that results are more accurate than manual segmentations of a second observer compared to ground truth segmentations. However, scoring does not consider the locations of errors. While ground truth segmentations follow a protocol for including parts of the vena cava into the segmentation, the second observer did not follow this protocol. We encoded this protocol into our automated segmentation method, which explains its superior scoring as compared to the second observer.

Figure 7.3 shows slices of three exemplary test cases (cf. Table 7.1), namely a relatively easy case (test case 1, top), an average case (test case 4, middle), and a relatively difficult case (test case 3, bottom). Difficulty was rated by the MICCAI workshop *3D Segmentation in the Clinic: A Grand Challenge* organizers according to average performance of all competing automatic segmentation systems. The consideration of tumors inside the liver during the segmentation process (see Sec. 7.1.3) is important and works well on the evaluated data (see Fig. 7.3 middle and bottom). Automated segmentations can exhibit errors in regions where anatomical structures with very similar intensity values are located close to the liver. Errors occur particularly if parts of such structures can be captured within a typical liver shape. This can be the case for the lower part of the vena cava (Fig. 7.3 top, right), the duodenum (Fig. 7.3 middle, right), the heart, muscles, stomach and pancreas. Some deviations are caused by incorrect manual segmentations (e.g. Fig. 7.3 middle row, middle column, near the gallbladder). Furthermore, the visibility problem stem-

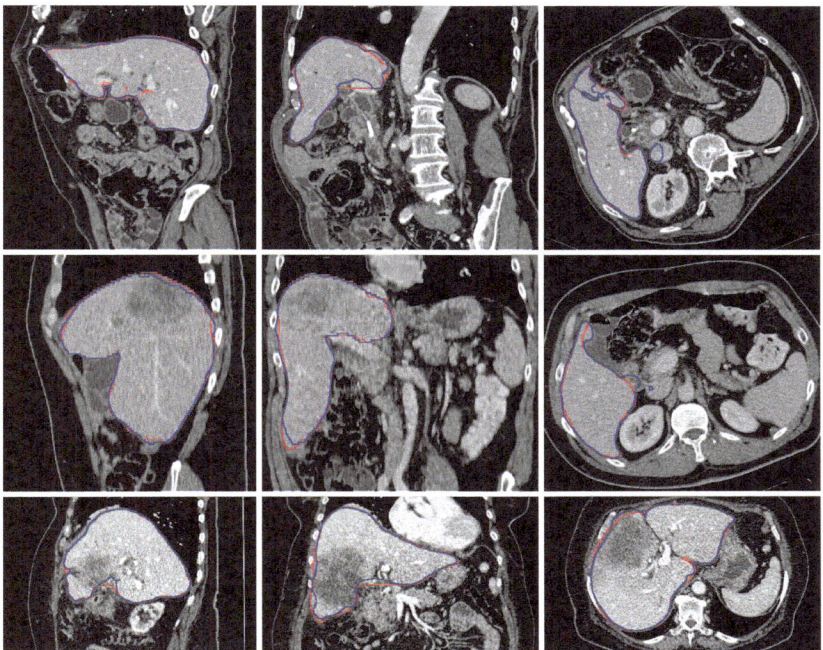

Figure 7.3 From left to right, a sagittal, coronal and transversal slice from three test cases. Red contour: Gold standard segmentation. Blue contour: Automatic segmentation as described in this section. Slices are displayed with an intensity window of $[-130, 270]$ HU.

ming from unidirectional intensity profiles as explained in Chapter 4 causes inaccuracies at the ridge-shaped, highly curved transition region from anterior/proximal to posterior/distal side of the liver (Fig. 7.4).

In summary, in this section we have presented a conceptually simple approach for fully automatic segmentation of the liver in CT. Our approach outperforms related work on benchmark image data in terms of accuracy. To allow for the application of outlier-resistant free deformation models like GraphCuts and ODDS (see Section 3.4.2 and Chapter 4), future work will focus on establishing a non-binary appearance cost function that does not directly select a single candidate displacement. In the same go we may be able to omit image filtering and hence eliminate the risk of "filtering away" weak yet relevant image features. We hypothesize that a combination of the flexibility as provided by the FreeBand approach and robustness

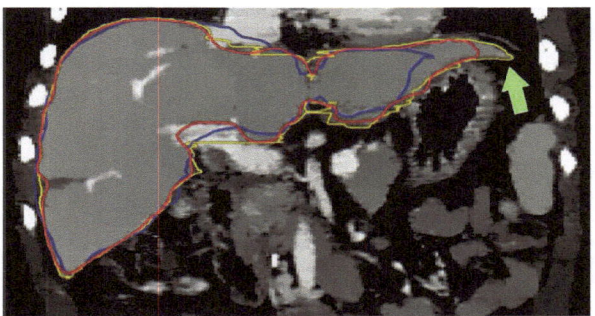

Figure 7.4 Visibility problem in exemplary coronal slice of liver CT. Contours: Yellow: Gold standard segmentation. Blue: Deformed SSM. Red: FreeBand result. Neither SSM deformation nor FreeBand provide for an accurate segmentation of a ridge-shaped region of the liver as indicated by green arrow. ODDS (cf. Chapter 4) may improve accuracy, yet require a non-binary appearance cost function, which is subject to future work.

of the GraphCuts approach will improve accuracy in regions of faint image contrast but strong shape knowledge, particularly in regions where the liver resides adjacent to muscle tissue between ribs. Furthermore, the application of ODDS instead of FreeBand may improve segmentation in the ridge-shaped transition region from anterior/proximal to posterior/distal side of the liver.

7.2   Segmentation of the Pelvic Bones in CT

In this section we present an algorithm for fully automatic segmentation of the pelvic bones from CT that builds upon the work of Lamecker et al. (2004a). We propose a combination of the Generalized Hough Transform (Section 3.1.4), segmentation based on a statistical shape model (Section 3.1) and a free form segmentation step based on optimal graph searching (Section 3.4.2). An evaluation on 50 CTs acquired after unilateral total hip arthroplasty (THA) reveals high segmentation accuracy despite low resolution and heavy metal artifacts induced by hip joint implants. In terms of accuracy, our results as presented in Section 7.2.3 outperform segmentation results of related segmentation methods for which respective accuracy measures were assessed. However, even if analogous error measures are assessed, results are not directly comparable because evaluations are performed on different data pools. No benchmark set of pelvic CTs is available for comparative evaluation.

As for related work, Yokota et al. (2009) present an SSM-based segmentation method and report an average mean surface distance of 1.2 mm in a quantitative evaluation on 44 hip bones (i.e. without sacrum) in 22 pre-THA CTs of unspecified resolution. Ehrhardt et al. (2004) present an approach that employs *atlases* for segmentation of the pelvic bones from CT via non-linear image registration (cf. Sec. 5.2). As accuracy measure *sensitivity* compared to manual reference segmentations was reported for a pool of six CTs: 98.5% of gold standard bone voxels were reported to be labeled correctly. Neither error statistics nor alternative error measures are reported. Approaches without quantitative evaluation include the work of Pettersson et al. (2006) who register a prototype atlas containing manually segmented femur and pelvis represented by few gray values to patient data, the works of Haas et al. (2008) and Vasilache and Najarian (2008) who both propose a complex combination of low-level techniques, and the work of Wu et al. (2011) who propose an SSM-based approach.

A scheme for creating and validating volumetric statistical shape models of bones was presented by Chintalapani et al. (2007). Taking the male pelvis as an example, a statistical shape model was created from 110 CT datasets. Reduced shape models consisting of 90 training shapes were shown to approximate left-out training shape surfaces with an average surface distance of 1.5 mm. The shape model was also used for segmentation, but no evaluation w.r.t. ground truth was reported.

7.2.1 Image Data and SSM of the Pelvic Bones

For our work 50 CTs were available stemming from a clinical study that aimed at determining the long-term clinical outcome of uni-lateral THA.[1] The database is composed of half female and half male pelvises. With a voxel size of about $0.9 \times 0.9 \times 5 \, \text{mm}^3$ all CTs approximately have the same resolution. The CTs were manually labeled by students well trained in pelvic anatomy and image segmentation. These segmentations include the entire pelvis with all three adjoining bones, namely the left and right hip bone as well as the sacrum including the coccyx. Each hip bone consists of ischium, ilium, and pubis. At the prosthesis side the implant itself was not included in the reference segmentation. In order to allow for an independent evaluation of native and implanted side, datasets were mirrored before further processing such that all had the native hip joint on the right and the implant on the left side.

The statistical shape model used for segmentation was created (as described in Sections 3.1.1 and 3.1.2) from the above mentioned manual segmentations of 50 pelvic CTs. Hence the model meets the recommendation of Chintalapani et al. (2007), according to which 40 - 50 training shapes are at least necessary to capture shape variations of the pelvis. The SSM is a non-manifold surface mesh (cf. Sec.

---

[1] Thanks to Markus Heller (Julius Wolff Institute, Charité - Universitätsmedizin Berlin, Germany) for providing image data of the pelvis. Thanks to Alexander Wurl and Philippe Moewis (Julius Wolff Institute) for manually segmenting the pelvic bones.

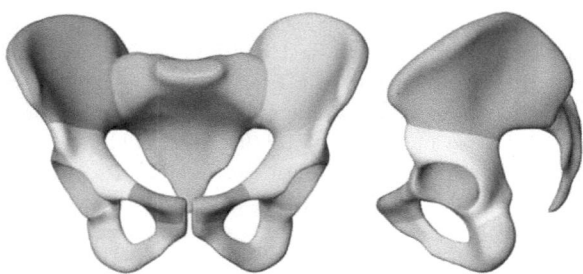

Figure 7.5 Mean shape $\bar{v}$ of the statistical shape model created from 50 pelvises. Colors/gray levels indicate the patch structure used for establishing point-to-point correspondences between training shapes.

2.1.3) containing two *inner* surfaces separating the sacrum from the adjoining hip bones. It consists of 29619 vertices and 59403 triangles, divided into 21 patches (see Figure 7.5).

7.2.2   Segmentation Pipeline

Our segmentation pipeline consists of a series of steps as described in Algorithm 8, namely pose initialization (**GHT**, see Sec. 3.1.4), SSM deformation (**SSM**, see Sec. 3.1.3), and free deformation via GraphCuts (**GC**, see Sec. 3.4.2). Parameters that are specific for each step are specified in the algorithm.

As appearance cost function we employ the continuous cost function ϕ described in Section 3.2.1. The boundary between bone and surrounding soft tissue appears as a transition from bright (bone) to darker (soft tissue) intensities. Hence we expect negative directional derivatives along bone surface normals, and set the respective condition for low appearance cost to $\nabla_{n_v} I(v + s) < -g < 0$. As for specific values of appearance parameters, we set $[t_1, t_2] := [120, 320]$ as intensity window, and $g := 50/\text{mm}$ as gradient threshold. Note that cortical bone usually exhibits higher intensities in CT, namely more than 500 Hounsfield Units (see e.g. Kalender (2011)). We set a "darker" intensity window following the observation that partial volume effects cause considerably lower intensities at the bone surface.

7.2.3   Results and Discussion

50 CTs as described in Section 7.2.1 were segmented automatically applying the algorithm presented in Section 7.2.2. As usual we employ the SSM in a leave-one-out manner.

Algorithm 8 Pelvis segmentation. Parameters stay the same unless noted otherwise.

1: **GHT:** Image re-sampling to voxel size $5 \times 5 \times 5 \, \mathrm{mm}^3$, scale range $[0.8, 1.2]$, rotation range $\pm 10°$, ignore locations with intensities below $100 \, \mathrm{HU}$.
2: **SSM** with profile length $2r := 50 \, \mathrm{mm}$, number of sample points $n := 51$.
3: **SSM** with $2r := 20 \, \mathrm{mm}$
4: **GC** with $2r := 20 \, \mathrm{mm}$, $n := 41$, shape preservation $c := 1 \, \mathrm{mm}$

After each step (pose initialization, SSM deformation and GraphCuts) the resulting surface meshes were converted to label images with the same resolution as the respective CT. The non-manifold structure of the SSM as described in Section 7.2.1 allows for a separate conversion of each sub-bone, yielding individual labels for the right hip bone, the left hip bone and the sacrum. Manual reference segmentations assign individual labels for these structures, too. Hence quantitative evaluation of single labels as well as any combination of labels is possible. Note that the evaluation of a combination of labels does not take into account the boundaries between the respective sub-bones. For evaluation, we used the following metrics: Relative volume difference (RVD), absolute value of RVD (ARVD), Dice overlap error (DOE), symmetric mean, root mean square and Hausdorff surface distance (MD, RMSD, HD), each as described in Section 6.1.

Results are given in Table 7.2. The average (bold) and standard deviation (in brackets) of each error metric as assessed for 50 test cases is shown. Results for the whole pelvis (All), the right hip bone (RHB), the left hip bone (LHB) and the sacrum (S) are listed in separate rows. Table 7.2 lists the evaluation results for *intermediate* segmentations after GHT and SSM deformation of the pelvis (top and middle table, respectively), and evaluation results for *final* segmentations after free deformation with GraphCuts (bottom table). Additionally, result statistics are presented as box plots in Figure 7.6, where the median (black diamond), the inter-quartile range (colored boxes), and outlier-free min-max ranges are visualized for all metrics and all segmentation phases.

Final segmentations of the overall pelvis exhibit an average MD of $0.6 \pm 0.3 \, \mathrm{mm}$, an average RMSD of $1.7 \pm 0.8 \, \mathrm{mm}$, and an average HD of $16.1 \pm 5.9 \, \mathrm{mm}$. The average volumetric errors are $-3.0 \pm 3.8\%$ for the RVD, $3.7 \pm 3.1\%$ for the ARVD and $12.8 \pm 3.3\%$ for the DOE. Figure 7.7 shows three final segmentation results that are exemplary for a good, an average and a bad case in terms of their MD.

Final segmentation accuracies of the three anatomical structures, namely left hip bone, right hip bone and sacrum, exhibit significant differences. The right hip bone, with an average MD of $0.3 \pm 0.1 \, \mathrm{mm}$, is segmented more accurately than its left counterpart reaching $0.6 \pm 0.2 \, \mathrm{mm}$ (see Table 7.2, bottom). Both hip bones and the whole pelvis reach an average MD smaller or equal to the in-plane resolution of the

GHT	RVD [%]	ARVD [%]	DOE [%]	MD [mm]	RMSD [mm]	HD [mm]
All	0.8 (16.2)	12.1 (10.7)	49.0 (8.3)	3.8 (1.3)	5.4 (1.8)	24.1 (6.5)
RHB	1.0 (16.8)	12.6 (10.9)	50.6 (8.7)	3.4 (1.2)	4.7 (1.6)	18.6 (4.3)
LHB	1.7 (19.2)	14.5 (12.5)	53.5 (14.0)	3.8 (2.2)	5.3 (2.7)	20.3 (7.1)
S	0.7 (16.4)	12.3 (10.8)	48.0 (13.6)	4.5 (2.1)	5.8 (2.4)	19.7 (5.9)
SSM	RVD [%]	ARVD [%]	DOE [%]	MD [mm]	RMSD [mm]	HD [mm]
All	-3.9 (5.3)	5.0 (4.2)	21.8 (3.6)	1.2 (0.3)	2.2 (0.6)	16.1 (5.4)
RHB	-5.1 (4.5)	5.7 (3.6)	20.8 (2.7)	0.9 (0.2)	1.5 (0.2)	9.2 (1.9)
LHB	-4.6 (6.4)	6.5 (4.5)	24.4 (4.1)	1.0 (0.2)	1.9 (0.3)	11.4 (2.0)
S	-0.8 (11.5)	8.0 (8.3)	25.9 (6.0)	1.8 (0.7)	3.0 (1.2)	15.6 (5.7)
GC	RVD [%]	ARVD [%]	DOE [%]	MD [mm]	RMSD [mm]	HD [mm]
All	-3.0 (3.8)	3.7 (3.1)	12.8 (3.3)	0.6 (0.3)	1.7 (0.8)	16.1 (5.9)
RHB	-3.8 (2.5)	3.9 (2.4)	9.7 (2.0)	0.3 (0.1)	0.8 (0.2)	7.7 (2.1)
LHB	-2.7 (4.9)	4.1 (3.8)	14.8 (3.7)	0.6 (0.2)	1.5 (0.4)	11.3 (2.4)
S	-1.7 (11.4)	7.2 (8.9)	18.6 (6.4)	1.2 (0.7)	2.5 (1.5)	15.2 (6.3)

Table 7.2 Evaluation results for initialization with GHT (top), SSM deformation (middle) and free deformation with GraphCuts (GC, bottom). Average error measures with standard deviations in brackets.

CT data. Compared to the hip bones, the average MD for sacrum segmentations is relatively large with 1.2 ± 0.7 mm.

Among all 50 cases, the largest MD after pose initialization with the GHT is 6.3 mm. The evolution of error metrics over the three segmentation phases (see Figure 7.6) shows that error metrics are improved in each phase, with the only exception that the average HD does not change considerably from SSM deformation to final segmentation result.

Discussion

The GHT proved to be a robust method to estimate the pose and size of the pelvis in CT. We attribute the fact that the pelvis, despite its varying anatomy, can be detected by only a single average template shape to its unique shape within the human anatomy and to the distinct image features it exhibits in CT.

Figure 7.7 explains differences in segmentation accuracy between right and left hip bone: Mesh deformation is likely to pull towards the outer contour of implants due to strong image features there, whereas manual segmentation left out the implant, leading to larger errors at the implanted acetabulum.

Another problem at the acetabulum which concerns pathological *and* healthy side is *leakage* of the segmentation into the femoral head due to a lack of image features (see Figure 7.8). However, extreme leakage is confined by the shape preserving

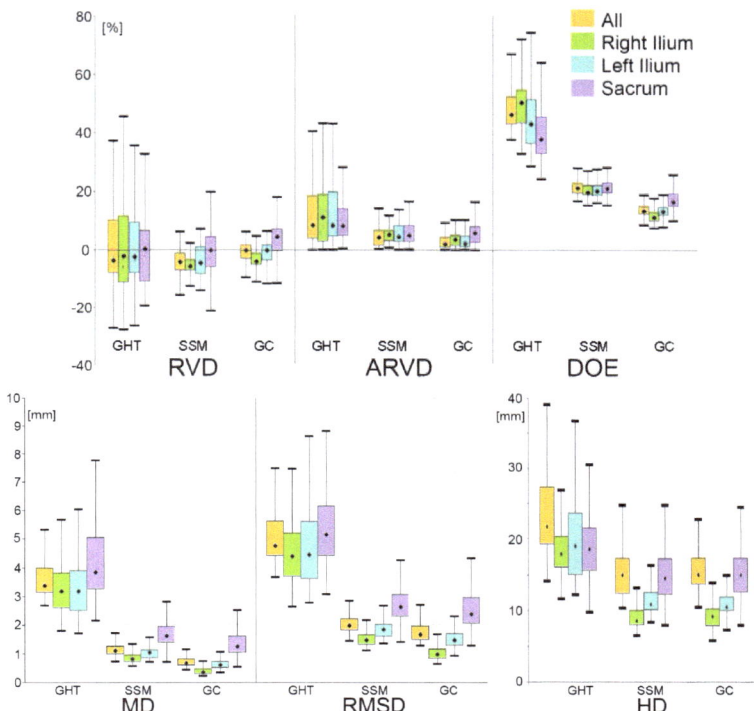

Figure 7.6 Volumetric and surface distance error measures after GHT, SSM deformation, and GraphCuts (GC).

constraint of GraphCuts. Multi-object GraphCuts (cf. Sec. 3.5) are able to further constrain leakage and yield more accurate results for the acetabulum, as evaluated in Section 8.1.

The comparatively large error metrics for the sacrum are caused in large part by an over-segmentation of the lower lumbar vertebra due to the lack of decisive image information in this area (see Figure 7.9a,b). Furthermore, on the tip-like coccyx (i.e. the tailbone of the sacrum), the visibility problem for normals occurs (cf. Sec. 4.1), which can be remedied with *ODDS* (cf. Sec. 4.2), as illustrated for an exemplary case in Fig. 7.9c. Since the sacrum is the smallest of the three structures in terms of surface area and volume, these errors have a rather small impact on the overall result, especially when disregarding the boundaries to right and left hip bone.

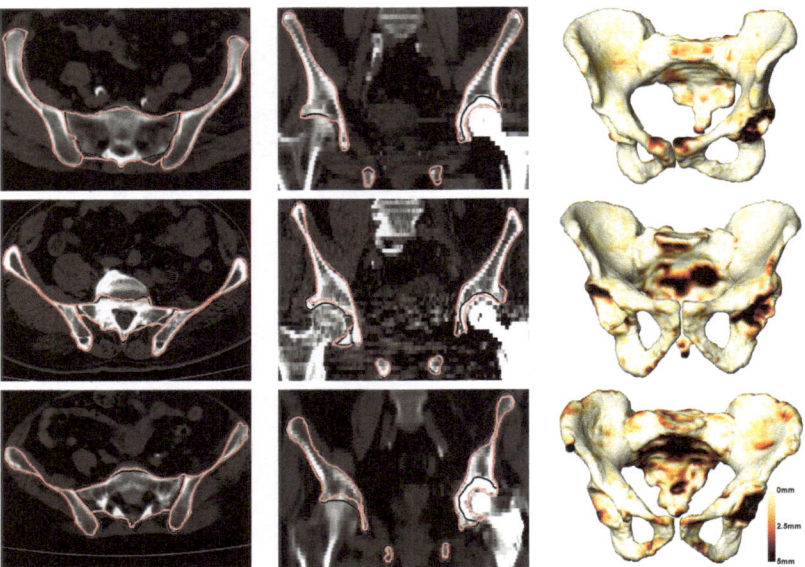

Figure 7.7 Exemplary pelvic bone segmentations. From left to right: Transversal, coronal and 3D-view of a good case (top, MD 0.5 mm), an average case (middle, MD 0.7 mm) and a difficult case (bottom, MD 1.2 mm). Manual segmentations outlined in black; automatic segmentations outlined in red. 3D views show automatic segmentations, with surface distance to respective references encoded by color. Systematic inaccuracies at left acetabulum stem from gold standard segmentation protocol at implanted joint: Reference segmentations exclude implants, while automatic segmentations adapt to respective strong image features.

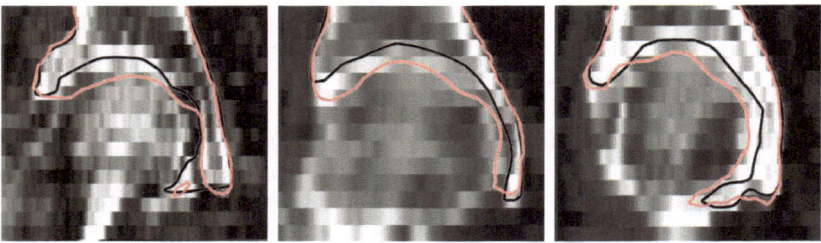

Figure 7.8 Acetabulum (i.e. hip joint of pelvic bones). Black: Gold standard. Red: GraphCuts. GraphCuts segmentation protrudes into the femoral head.

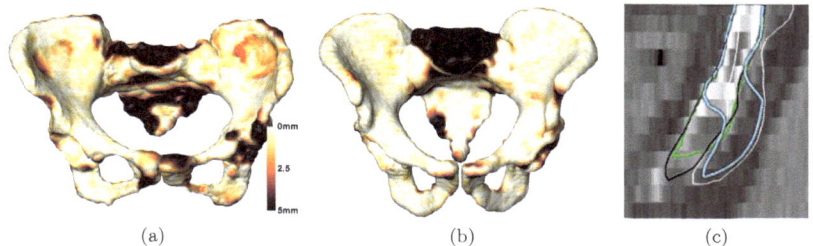

(a)   (b)   (c)

Figure 7.9 (a,b) Distance maps of automatic to reference segmentations. Exemplary cases with large errors at lower lumbar vertebra of the sacrum. (c) Coccyx (tailbone) of the sacrum. Exemplary case revealing the visibility problem that comes with unidirectional intensity profiles. Contours: Black: Gold standard. White: Initial mesh. GraphCuts result similar to initial mesh; not shown. Blue: FreeBand result. Green: *ODDS* as presented in Section 4.2 overcome the visibility problem.

A reason for the stagnation of the maximum distance values from SSM- to free deformations is the profile length $2r := 20\,\mathrm{mm}$ chosen for free deformation: HD values larger than $10\,\mathrm{mm}$ after SSM deformation implicate that there are vertices on the deformable mesh that are *too far away* from the target surface such that it cannot be reached along profiles. However, longer profiles result in higher errors, which may be attributed to more outliers.

In summary, in this section we have proposed a fully automatic pipeline for segmentation of the pelvic bones in CT. We have presented an evaluation of segmentation accuracy on 50 clinical CTs. Concerning improvements of segmentation accuracy, future work will establish a region-specific appearance model for the proximal end of the sacrum adjacent to the lower lumbar vertebra. To this end, it may be beneficial to include the lower lumbar vertebra into the SSM. Furthermore, the accuracy of SSM deformation needs to be improved such that the target object boundary is "within reach" before deformation with GraphCuts. To this end, we see potential in extending the training set of our SSM. Also, establishing SSMs of different "shape classes" (e.g. male/female) may be beneficial.

Evaluation results suggest that our approach outperforms related work in terms of segmentation accuracy. However, due to a lack of benchmark data, results are not directly comparable. Furthermore, we cannot put our results into context w.r.t. to manual segmentations due to a lack of information on inter-observer variability. Establishing a common pool of benchmark pelvic CT together with standard accuracy measures as well as assessing inter-observer variability of manual segmentations is indispensable for a more conclusive evaluation, and subject to future work.

7.3   Segmentation of the Mandibular Bone and Nerve in CBCT

Three-dimensional imaging has become an important technology for diagnosis and planning in dentistry and maxillofacial surgery (Schramm et al., 2005). Cone beam computed tomography (CBCT) presents an alternative to conventional CT because of its affordable costs as well as its reduced dose per examination. Thus, CBCT is likely to become a preferred imaging technique for dental practices. One major application for CBCT is dental implantology, where a primary concern is an optimal and stable placement of implants within the jaw bone without any impairment of the facial nerves. As a side effect of low dosage, however, the signal to noise ratio is not as high as for CT and soft tissue structures cannot be discriminated clearly. This renders the exact localization of the mandibular nerve canal within the bone highly challenging.

Stein et al. (1998) present a method for interactively segmenting the nerve canal in CT, yet report only qualitatively good agreement with gold standard segmentations on five datasets. Hanssen et al. (2004) suggest a level-set approach for interactive 3D segmentation of the nerve canals in CBCT, but do not present any quantitative validation. Rueda et al. (2006) propose a semi-automatic system to perform 2D segmentation of the lower cortical and trabecular bone as well as detect the nerve canal and center in specific 2D slices of conventional CT. Their method is based on an active appearance model, and requires manual initialization. It yields an average accuracy of 1.6 mm for the cortical bone and 3.4 mm for the dental nerve in 215 single 2D slices. Yau et al. (2008) propose a semi-automatic method to segment the nerve canal from conventional CT data. It requires the user to manually specify a seed point for a subsequent automatic adaptive region-growing approach in consecutive slices. However, no quantitative validation is presented.

In contrast to existing work, the method we propose in this section provides (1) segmentation of the complete mandibular bone surface, and (2) delineation of the course of the mandibular nerves, both in 3d and in a fully automatic manner. Our method operates on CBCT instead of conventional CT. Our approach yields an accuracy that considerably surpasses the 2D results of Rueda et al. (2006). It is based on a compound SSM of the bone surface and the course of the nerve, which extends the work of Zachow et al. (2006). In order to match the SSM to CBCT data we extend the work of Lamecker et al. (2006b) in two major ways: (1) We segment the mandibular bone surface in a fully automatic fashion based on the algorithm presented in Section 7.2, which we adapted to the image characteristics of mandibular CBCT. (2) The estimated course of the nerves as extrapolated from the compound SSM is refined by means of an image-based tracing algorithm that we tailor to the specific appearance characteristics of CBCT.

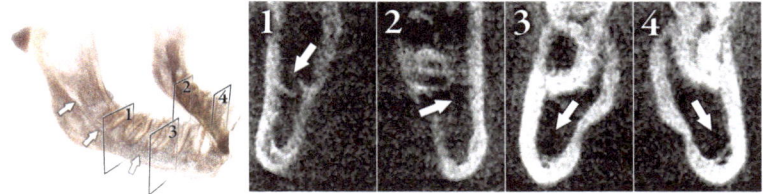

Figure 7.10 Volume rendering (left) and details of coronal slices (right) of CBCT data. Arrows indicate the location of the nerve canal.

7.3.1 Image Data and Compound SSM of Mandible and Nerves

Our work is based on 106 CBCTs of complete mandibles stemming from a PACS at University Hospital of Cologne, Germany.[2] All CBCTs were acquired with a Sirona Galileos CBCT at the maxillofacial surgery department (patients of age 16 to 71, 56 female, 50 male). CBCT is performed routinely in cases of suspected orbital floor fractures, mandibular condyle evaluation, wisdom teeth removal, abscesses, etc. Images are taken in seated position with a scan duration of about 15 seconds. All images consist of 512^3 voxels with an isotropic voxel size of 0.3 mm. The field of view is approximately $15\,cm^3$. The X-ray source is operated at 85 kV with a tube current of 5-7 mA. Fig. 7.10, right, depicts details of typical slices through such data. In each dataset the bone as well as the nerve canal were segmented manually by an experienced dentist.

The SSM used for automatic segmentation is generated on the basis of the manual segmentations described above. For each mandibular bone a triangular surface is generated. The mandibular surfaces are each divided into eight patches (Fig. 7.11) that are bounded by characteristic feature lines, detectable on every mandible, and corresponding meshes are established as described in Section 3.1.2. The teeth are not considered in the SSM due to an individually varying dentition state. For each pair of nerve canals piecewise linear center lines are computed using a skeletonization algorithm (Sato et al., 2000). The nerve center lines are consistently re-sampled with a fixed number of equidistant points starting at the mental foramen. The compound SSM of bone and nerves is computed as described in Section 3.1.1. The bone surface and nerve center lines contain 8561 and 200 vertices, respectively.

7.3.2 SSM-Based Reconstruction of Bone and Nerve

The SSM based method for reconstruction of the mandibular bone and nerves in CBCT is composed of the following steps: (1) Pre-processing of the image data

[2] Thanks to Max Zinser (Universiätsklinikum Köln, Germany) for providing image data and manual reference segmentations of the mandible.

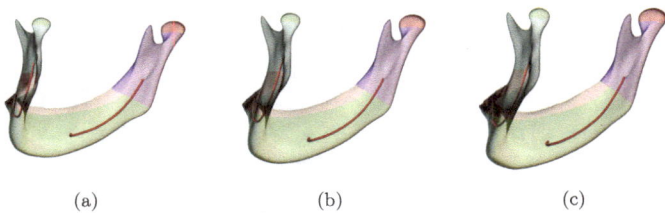

(a)   (b)   (c)

Figure 7.11 SSM of mandible bone and nerves: (b) Mean shape, (a/c) first mode of variation.

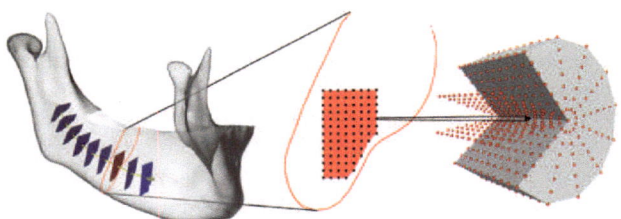

Figure 7.12 Dijkstra Optimization. (left) Normal planes along initial nerve reconstruction. (middle) Graph nodes on a normal plane. (right) Sampling cylinder at a graph node.

with a 3D median filter to cope with the increased noise level as compared to conventional CT, (2) position initialization, and (3) image driven deformation of the SSM described in Section 7.3.1. We employ the same framework (GHT, SSM deformation, appearance cost function) as for pelvis segmentation (cf. Sec. 7.2). Note that only mandibular bone mesh deformation is steered by appearance match, while the mandibular nerves are extrapolated from the SSM, as described in Section 3.1.5. We do not employ free deformations in this section, as the focus is on nerve delineation.   An evaluation of different free deformation methods on the mandibular surface is provided in Chapter 9.

7.3.3   Image-based Refinement of Nerve Delineation

SSM deformation as described in Section 7.3.2 yields fairly accurate segmentations of the mandible bone, as well as approximate nerve delineations. The SSM-based nerve delineations are not based on any image features, but are merely extrapolated by the SSM. We use them as initialization for image-based refinement: As for an appearance model, we assume that the nerve channel forms a dark tunnel through the bone which is partially surrounded by a brighter border. The key idea of our method is to build a graph through which the path with minimal cost from source to target corresponds to such a tunnel. To achieve this, we construct a graph with nodes weighted by a respective appearance cost function as described in the following. Note that all indices used for graph description start at 1, unless stated otherwise.

Graph nodes: For each vertex v_k at index k of the piecewise linear initial nerve representation, equidistantly distributed points in the normal plane at v_k serve as graph nodes. Fig. 7.12 shows some exemplary normal planes (left) and a normal plane with graph nodes (middle). The normal plane at v_k is spanned by two directions perpendicular to the line tangent $\vec{t}_k$, namely $\vec{y}_k := \vec{t}_k \times \vec{x}_{data}$, where $\vec{x}_{data}$ is the x-axis of the image reference coordinate system, and $\vec{x}_k := \vec{y}_k \times \vec{t}_k$. A graph node is described by vertex index k and its indices i, j on the respective normal plane. Let N_x, N_y be the number of nodes and X, Y the lengths for which the normal plane is considered in x_k- and y_k- direction, respectively. The position of node (k, i, j) is then $\vec{p}_{k,i,j} := v_k + l_{i,j}$, with $l_{i,j} := (\frac{i-1}{N_x-1} - 0.5) \cdot X \cdot \vec{x}_k + (\frac{j-1}{N_y-1} - 0.5) \cdot Y \cdot \vec{y}_k$. In addition to these nodes, two "artificial" nodes serve as source and target of the graph.

Graph edges: The graph contains directional edges from each node (k, i, j) to all nodes $(k + 1, i + di, j + dj)$ with $di, dj \in \{-1, 0, 1\}$, as well as directional edges from the source to all nodes with $k = 1$, and from all nodes with $k = N$ to the target, where N is the number of points on the nerve representation. This selection of edges limits differences of lengths of neighboring displacements to $\sqrt{2}\delta_L$, where δ_L is the sampling distance of the set of candidate displacements, $L := \{l_{i,j}\}_{i,j=1}^{N_x,N_y}$.

Node weights: For any node (k, i, j), a scalar cost function $\phi(v_k, l_{i,j})$ serves as weight. The cost $\phi(v_k, l_{i,j})$ is computed from image intensities sampled inside a cylinder with center $\vec{p}_{k,i,j}$, orientation $\vec{t}_k$, some length H and radius R. Fig. 7.12 (right) shows an exemplary cylinder with its sample points. A sample point is described by a length index h, a radius index r, and an angle index a. Let N_h be the number of sample points in length direction, N_r the number of sample points along a radius, and N_a the number of angles for which radii are sampled. Then the position of sampling point (h, r, a) is $\vec{p}_0 + \frac{h-1}{N_h-1} \cdot H \cdot \vec{t}_k + \frac{r-1}{N_r-1} \cdot R \cdot \vec{r}_a$ with $\vec{r}_a$ being a normalized radius vector, rotated by an angle $\frac{a-1}{N_a-1} * 2\pi$ around $\vec{t}_k$, and $\vec{p}_0 := \vec{p}_{k,i,j} - 0.5 \cdot H \cdot \vec{t}_k$. Note that for $r = 1$, no angle index is necessary to describe the sample point. To determine the cost $\phi(v_k, l_{i,j})$, the unfiltered image intensities at the cylinder sample points are evaluated as follows: The mean "inner" intensity μ^{inner} and standard deviation σ^{inner} is computed from all sample points located at radii no bigger than an "inner radius index" r_i. For each angle index a, the mean "border" intensity $\mu^{border}(a)$ and standard deviation $\sigma^{border}(a)$ is computed from all sample points at radii bigger than r_i and no bigger than a "border radius index" r_b. Furthermore, the mean "outside" intensity $\mu^{outside}(a)$ is computed for each angle index a from all sample points at radii bigger than r_b. The number n_a of angle indices is counted for which

$$\mu^{border}(a) - 0.1 \cdot \sigma^{border}(a) > \mu^{inner} + \sigma^{inner} \text{ , and}$$

$$\mu^{border}(a) - 0.1 \cdot \sigma^{border}(a) > \mu^{outside}(a) \text{ .}$$

If these conditions hold the average border intensity at angle a is considerably brighter than the average "inside" intensity of the cylinder, and also brighter than the average intensity beyond the border at angle a. This corresponds to a bright "rim" delimiting a dark "tunnel" through the cylinder, which is what we consider a feature of the nerve channel. The cost is then defined as $\phi(v_k, l_{i,j}) := \mu^{inner} - 50 \cdot n_a$. It is low if the cylinder has a dark core, and even lower if this core is surrounded by a bright rim which is in turn surrounded by a dark "outside". If a graph node position $\vec{p}_{k,i,j}$ lies outside the mandibular bone as reconstructed by the SSM, the respective cost is set to infinity (more precisely to some cost higher than the cost of any path through our graph that does not pass an infinite-cost node).

We compute the minimum cost path through the above described graph by means of Dijkstra's algorithm (Dijkstra, 1959), yielding our refined nerve delineation.

7.3.4   Results and Discussion

We evaluated SSM based bone segmentations and nerve delineations as well as nerve delineations refined by optimal path search on 106 CBCT datasets as described in Sec. 7.3.1. For each dataset, before performing SSM deformation, the respective training shape was removed from the SSM, i.e. the evaluation was conducted in a leave-one-out manner.

From data observation and experiments, we set parameters to the following values: SSM deformation: consider 80 shape modes, profile length $2r := 6$ mm, number of sample points $n := 51$. Bone appearance parameters $t_1 := 350$, $t_2 := 1250$, $g := 150/$mm. Optimization, graph nodes: $X := 12$ mm, $Y := 7$ mm, $N_x := 121$, $N_y := 71$, cylinder: $H := 3$ mm, $R := 2.1$ mm, $N_h := 11$, $N_r := 8$, $N_a := 12$, $r_i := 4$, $r_b := 6$. We determined these values empirically.

The average errors for mandible segmentations as compared to manual reference segmentations are: symmetric mean, root mean square and maximum surface distance: 0.5 ± 0.1 mm, 0.8 ± 0.2 mm, and 6.2 ± 2.3 mm, see also Figure 7.13a. Figure 7.13b shows exemplary bone reconstructions and their distances to the respective gold standard surface. Apart from errors around the teeth, the largest errors occur at the mental protuberance and the condyles. These errors are due to the increasing noise towards the fringe of the field of view. The average symmetric

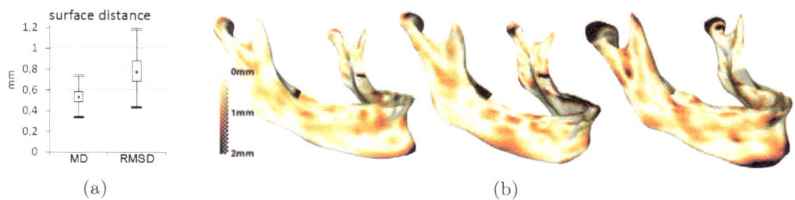

Figure 7.13 Surface reconstruction with statistical shape model. (a) Symmetric mean and root mean square surface distance errors (MD, RMSD). (b) From left to right: good, average and bad case. Color encodes distance to gold standard surface.

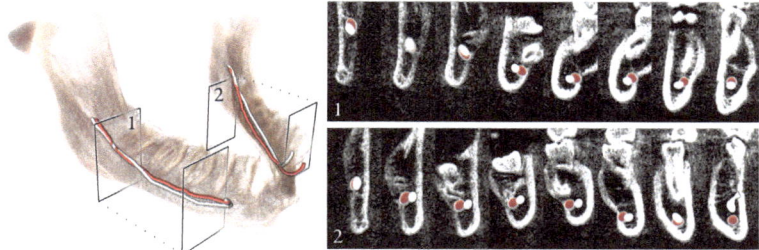

Figure 7.14 Image based nerve delineation via optimal path search: Exemplary bad case. White line: gold standard nerve. Red line: automatic nerve delineation. Error measures for this case: Mean distance to gold standard nerve: right 1.5 mm, left 2.1 mm. Fraction that lies within the gold standard nerve canal: right 45%, left 30%. Largest error at mental foramen (i.e. anterior end) of left nerve canal.

mean curve distance of SSM based nerve delineations to the respective ground truth is 1.7 ± 0.7 mm (right nerve), and 2.0 ± 0.8 mm (left nerve).

For nerve delineations after image-based refinement, the average symmetric mean curve distance to the respective ground truth is 1.0 ± 0.6 mm (right nerve), and 1.2 ± 0.9 mm (left nerve). The average fraction of the nerve that lies within the gold standard nerve canal is $80 \pm 24\%$ (right nerve), and $74 \pm 27\%$ (left nerve), see also Figure 7.15a. Figure 7.14 shows an exemplary nerve delineation with high error. Figure 7.15b shows the average *asymmetric* curve distance from SSM based and refined nerve delineations to gold standard nerve delineations per point along the curve. This illustrates that image-based refinement via optimal path search is able to reduce the delineation error considerably in a region in the middle of each nerve, while the reduction is not that obvious towards the ends of each nerve.

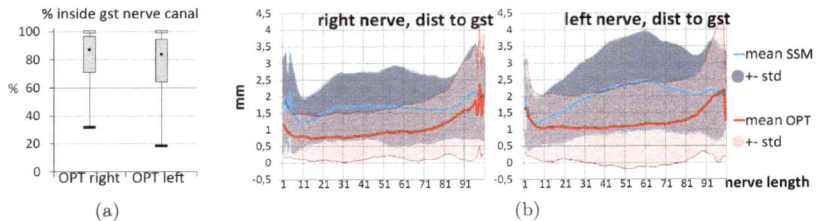

(a)   (b)

Figure 7.15 Evaluation of nerve delineation on 106 CBCTs. SSM-estimated initial delineation (SSM) and refined delineation resulting from optimal path search (OPT). (a) OPT: Box plot of nerve fraction within gold standard nerve canal. (b) Distance of initial and refined nerve delineations to gold standard (gst) along the nerve from posterior to anterior end (1..100).

In summary, in this section we have presented an accurate and robust method to automatically segment the mandibular bone and delineate the alveolar nerves in CBCT data. Future work will focus on improving the accuracy of nerve delineation, especially concerning the ends of the nerve canal. In this work we chose a conceptually simple tracing approach for nerve delineation. Other methods for tracing tubular structures may be considered as alternatives, e.g. as described for vessel detection (see e.g. Wong and Chung (2007); Poon et al. (2007) and references therein).

Note that we proposed a conceptually similar fully automatic pipeline for detection and identification of teeth in CT (Nguyen et al., 2012). To this end, future work will cope with accurate segmentation of individual teeth, which may in turn improve the accuracy of bone segmentation and nerve delineation.

7.4 Conclusion

In this chapter we have tailored pipelines that perform fully automatic single-object segmentation of anatomical structures for applications that are relevant in clinical practice. Pipelines are assembled from the toolkit described in Chapter 3. Evaluations on clinical data reveal that our segmentation results are highly accurate as compared to related work, as far as a direct comparison is possible (cf. Chapter 6).

However, concerning the pelvic bones and particularly the acetabulum, we have identified inaccuracies that we attribute to weak or ambiguous image features in a multi-object environment. Furthermore, concerning ridge-like structures on the liver and the pelvic bones, we have identified inaccuracies that we attribute to the *visibility problem* as described in Section 4.1. The remedial effect of respective designated tools, namely multi-object GraphCuts on coupled meshes (cf. Sec. 3.5) and ODDS/fastODDS (cf. Chapter 4), are subject to thorough evaluation in Chapters 8 and 9.

Chapter 8

Multi-object Segmentation of Joints

Contents

In Section 3.5 we described *multi-object GraphCuts* and proposed a *mesh-coupling algorithm*. Together, these methods allow for simultaneous segmentation of arbitrarily shaped adjacent objects. This chapter presents applications of coupled multi-object GraphCuts to selected structures of the musculoskeletal system in joint regions, namely bones and cartilage.

In Section 7.2 we achieved accurate segmentations of the pelvis based on an SSM and single-object GraphCuts (Section 3.4.2). However, inaccuracies occurred in the acetabulum, as no knowledge about the femur was exploited. In Section 8.1 we present a comparative evaluation of single- vs. coupled multi-object GraphCuts for segmentation of 50 pelvic CTs with focus on accuracy in the hip joint, as published in Kainmueller et al. (2009b,d). It shows that coupled multi-object GraphCuts outperform single-object GraphCuts in terms of segmentation accuracy in the hip joint.

Section 8.2 proposes a multi-object pipeline for fully automatic segmentation of knee bones and cartilages in MRI, as published in Seim et al. (2010). It is

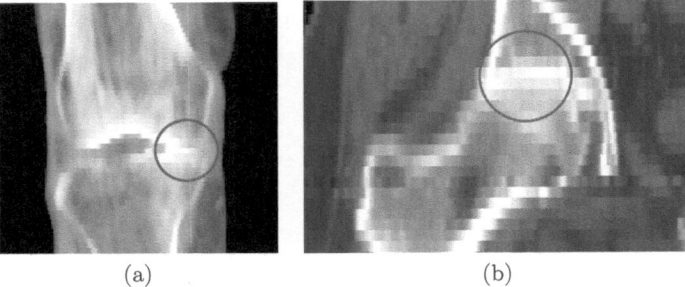

(a)   (b)

Figure 8.1 (a) CT of distal femur and proximal tibia, slice thickness 2 mm. (b) Acetabulum and proximal femur, slice thickness 4.6 mm. The joint space is barely visible in encircled areas. Nevertheless, segmentations of adjacent bones need to be *consistent*, i.e. non-overlapping; they need to be in accordance with image features as far as present, and they need to have plausible shapes.

assembled from the model kit described in Chapter 3 including coupled multi-object GraphCuts, extended by application-specific components. With this framework, we won the second prize at the MICCAI *Medical Image Analysis for the Clinic – A Grand Challenge* on-site knee bone and cartilage segmentation contest in 2010 (see www.ski10.org). As at June 2014 our method is the second-most accurate among all competing automatic methods in terms of a *score* (Heimann et al., 2010) that measures segmentation accuracy on benchmark data.

As before we do not focus on the run-time of segmentation pipelines in this chapter. However, note that average run-times are five minutes for the hip joint and six minutes for the knee on a 3.2 GHz Core.

8.1   Segmentation of the Hip Joint in CT Data

For patient-specific biomechanical simulations, e.g. of the human lower limb, an accurate reconstruction of the bony anatomy from medical image data is required. This particularly applies to joint regions, as simulation results heavily depend on the anatomy of joints (Heller et al., 2001). Bone appears well-contrasted in CT as compared to other imaging modalities, which makes CT widely used for joint replacement planning (see e.g. Bui and Taira (2010)).

In CT data, bony tissue usually exhibits high intensity contrast to surrounding soft tissues. However, in joint regions, single-object segmentation techniques are often not sufficient for accurate separation of adjacent individual bones. Due to large slice thickness or pathological changes of bones, the joint space may be hard to detect even for human observers. Figure 8.1 shows exemplary situations.

The objective of this section is accurate and consistent (i.e. non-overlapping) segmentation of femur and pelvis in CT data. The focus is on accuracy in the hip joint. Apart from interactive and semi-automated approaches (Liu et al., 2008; Kang et al., 2003), Yokota et al. (2009) present a fully automatic segmentation method and report an average mean surface distance of 1.8 mm in the region around the joint space (on both sides) in a quantitative evaluation on 44 hip bones in 22 pre-THA CTs of unspecified resolution. Furthermore, Zoroofi et al. (2003) evaluate a series of low level techniques for fully automatic segmentation of the femoral head and acetabulum in 60 pre- and post THA CTs of unspecified resolution w.r.t. custom quality measures. In terms of accuracy, the segmentation pipeline presented in this section outperforms related segmentation methods for which respective accuracy measures were assessed. However, as discussed in Section 7.2, comparable error measures do not provide for comparable results as evaluations are performed on different data pools due to a lack of benchmark hip CTs.

The segmentation pipeline proposed in this section is composed of the following steps: (1) Initialization by detection of the pelvis (cf. Sec. 7.2), (2) Deformation of an *articulated* SSM (ASSM) of femur and pelvis as proposed in Kainmueller et al. (2009b)[1], and (3) Coupled multi-object GraphCuts. We employ the appearance model and cost function as for pelvis segmentation (cf. Sec. 7.2.2), with the same parameters.

As for the parameters of mesh-coupling and multi-object GraphCuts, from experiments, we set the minimum angle that shared displacement directions are allowed to have to surfaces to $\alpha := 30°$. We found this angle to be a reasonable compromise between large enough sizes of coupled regions (ideally covering the whole acetabulum on the pelvis side) which calls for a smaller angle, and consistency of displacement directions between coupled and non-coupled regions which calls for a larger angle. Furthermore, we set minimum and maximum distance constraints to $c_0 := 0$ and $c_1 := 8$ mm, respectively. A minimum distance of zero accounts for very small joint spaces of arthritic hip joints that appear in our database. A maximum distance as large as 8 mm is intended to avoid too tight constraints towards the boundary of the coupled region, which we assume to correspond to the acetabular rim on the pelvis side of the joint.

We evaluate segmentations of femur and pelvis in the 50 CTs described in Section 7.2.1. All CTs display the pelvis, femoral heads, and various amounts of the femoral body. At maximum, about half the femoral body is displayed. In addition to the pelvis, the right proximal femur was segmented manually to provide gold standard segmentations.

We perform a comparative evaluation w.r.t. a second set of automatic segmenta-

---

[1]Kainmueller et al. (2009b) present preliminary work on ASSMs that has been extended by Bindernagel et al. (2011); Bindernagel (2013).

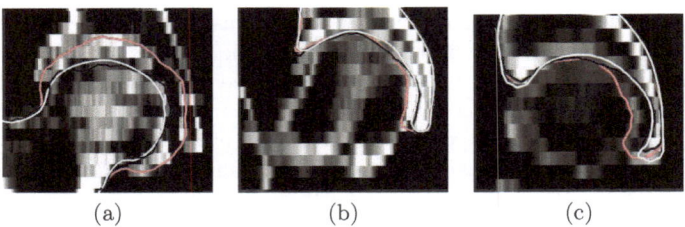

(a)   (b)   (c)

Figure 8.2 Black: Coupled multi-object GraphCuts. White: Gold standard. Red: Single-object GraphCuts. (a) Coupled multi-object GraphCuts prevent mis-adaptation of femoral head to features inside the pelvis. (b,c) Coupled multi-object GraphCuts prevent mis-adaptations of acetabulum via minimum distance constraint.

	ASSM		Single GC		Multi GC	
	md	hd	md	hd	md	hd
F	**1.7** (0.6)	**4.5** (1.5)	**1.7** (1.0)	**5.8** (2.1)	**1.1** (0.5)	**4.3** (1.4)
A	**1.4** (0.5)	**5.4** (1.9)	**0.9** (0.5)	**4.7** (2.4)	**0.8** (0.5)	**4.7** (2.2)

Table 8.1 Segmentation accuracy in the hip joint assessed for 50 CTs. Methods: Initialization (ASSM), single-object GraphCuts (Single GC) and coupled multi-object GraphCuts (Multi GC). Results for femoral head (F) and acetabulum (A). Bold: Average errors. Standard deviations in brackets. All entries in mm.

tions which were generated with *single-object* GraphCuts instead of coupled multi-object GraphCuts using the same parameters, meshes and intensity profiles.

The focus of our evaluation is to determine how accurate segmentations are in the area of the right hip joint. As accuracy measures we compute the mean and maximum surface distance (md, hd) of automatically determined right acetabulum and femoral head from the respective gold standard surface. Each structure is defined by a set of vertex indices which form *patches* on gold standard surface meshes. Thus we can compute the distance from these patches to the respective automatic segmentation. Note that consequently, md and hd are asymmetric measures. Computing respective distances in reverse direction, i.e. from automatic to manual segmentations, is technically more challenging because mesh coupling alters the patch structure of surfaces, and is subject to future work.

Table 8.1 lists the average error metrics for the femoral head and acetabulum. Results are given for the ASSM-initialized surfaces, single-object GraphCuts, and multi-object GraphCuts on coupled surfaces.

Discussion

Evaluation of segmentation accuracy on 50 CTs as summarized in Table 8.1 show that multi-object GraphCuts on coupled surface meshes of femur and pelvis yield more accurate segmentations of the hip joint than single-object GraphCuts on separate meshes. For the femoral head, coupled multi-object GraphCuts yield significantly more accurate results than single-object GraphCuts (average md 1.1 versus 1.7 mm). In fact, single-object GraphCuts do not improve initial segmentations at all, as the femur adapts to image features inside the pelvis, as shown in Figure 8.2a. Such mis-adaptations are prevented by the distance constraints in coupled multi-object GraphCuts. As for the acetabulum, coupled multi-object GraphCuts yield slightly – but not significantly – more accurate results than single-object Graph-Cuts (average md 0.8 versus 0.9 mm). Although most individual results are similar, coupled multi-object GraphCuts outperform single-object GraphCuts in a few test cases due to the distance constraints, as shown in Figure 8.2b,c.

With ASSM-based initialization, the problem of initially overlapping regions that cannot be mapped during mesh-coupling (cf. Sec. 3.5.2 and Fig. 3.4c (bottom)) did not occur for our 50 test cases. Consequently, femur and pelvis segmentation results do not overlap in any case.

In summary, multi-object GraphCuts on adjacent meshes coupled with shared displacement directions allow for consistent segmentation of the hip joint in CT with improved accuracy as compared to single-object GraphCuts. Average error measures suggest that our approach competes well with related work on hip joint segmentation in CT in terms of accuracy, yet results are not directly comparable due to different pools of test data.

Note, with a fully automatic approach for segmentation of the mandibular bone in CT which is conceptually similar to the pipeline proposed in this section, we won the first prize at MICCAI *3D Segmentation in the Clinic: A Grand Challenge* on-site mandible segmentation contest in 2009 (see Kainmueller et al. (2009c); Pekar et al. (2009)).

8.2 Segmentation of Knee Bones and Cartilage in MR Data

Osteoarthritis (OA) is a disease of articular cartilage which is estimated to be a leading cause of disability in more than 10% of the world population over the age of 60 (see e.g. Majumdar (2010)). Monitoring OA progression or drug response requires exact quantification of knee cartilage. Relevant measures include the area of bone-cartilage interface, cartilage thickness, and cartilage volume (Graichen et al., 2004). MRI is capable of imaging cartilage with good contrast as compared to other imaging modalities (see e.g. Bui and Taira (2010)).

In this section we propose a pipeline for fully automatic segmentation of knee

bones and cartilage from MRI. First, the femoral and tibial bone surfaces are reconstructed based on a pipeline employing GHT, SSM deformation and GraphCuts
(cf. Sec. 7.2). Starting from bone surfaces, cartilages are segmented simultaneously with coupled multi-object GraphCuts, exploiting prior knowledge about the
inter-individual variability of cartilage thickness.

As for related work, Folkesson et al. (2005) propose an automatic approach for
segmentation of articular cartilage based on supervised learning and report an average Dice coefficient of 0.80, sensitivity of 90.0% and specificity of 99.8% in an
evaluation on 46 MRIs with no or mild OA symptoms. Glocker et al. (2007) segment the cartilage of the patella with MRF-based registration of an atlas, achieving
an average Dice of 0.84, sensitivity 94.1%, specificity 99.9% and mean surface distance 0.49 mm on 28 MRIs from unspecified sources. Fripp et al. (2007) propose
an SSM-based approach and report average Dice coefficients of 0.96, 0.96 and 0.89
for femur, tibia and patella, respectively, and an average point-to-surface error of
0.16 mm on the bone-cartilage interface in an evaluation on 20 MRIs of healthy
persons. A multi-object GraphCuts based approach involving minor manual interaction ($\approx$ 30 sec) was proposed by Yin et al. (2009), where an evaluation on 16 MRI
datasets showed average mean surface errors of 0.2 to 0.3 mm for femoral and tibial
bones and 0.5 to 0.8 mm for cartilages.

Comparing the performance of these different approaches is difficult, because evaluation results strongly depend on varying properties and origins of the employed
image data (e.g. MRI sequences, varying types of pathologies) as well as different
evaluation metrics. To allow for direct comparability, Heimann et al. (2010) introduced a benchmark for automatic knee segmentation systems. We evaluate our
segmentation pipeline using the respective benchmark data pool consisting of 40
clinical MRIs and evaluation metrics as described by Heimann et al. (2010).

8.2.1   SSMs of Femur and Tibia and Cartilage Thickness Model

The statistical shape models of proximal tibia and distal femur used in this work
were generated from 60 MRI datasets provided by MICCAI 2010 workshop *Medical
Image Analysis for the Clinic – A Grand Challenge* (Heimann et al., 2010). For each
of these datasets a manual reference segmentation was available for the bones and
the associated cartilage. We fit existing SSMs of femur and tibia to gold-standard
segmentations, thus extrapolating the femoral and tibial shafts not included in the
field-of-view of the MRI datasets. We used the resulting reconstructed surfaces
to generate a new SSM for each bone covering the range of bone shaft portions
occurring in the 60 MRI datasets.

Bone training surfaces are equipped with corresponding meshes for SSM generation. Hence we can learn from training data the minimum, mean and maximum
cartilage thickness in normal direction per vertex of bone surface meshes. The

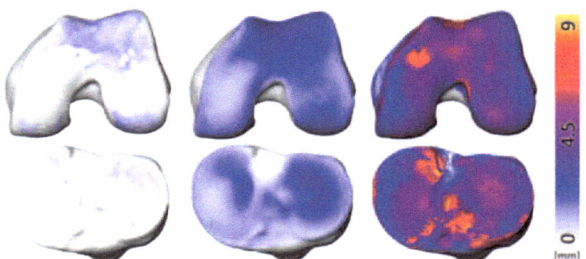

Figure 8.3 Cartilage: From left to right: Minimum (r_{min}), median and maximum (r_{max}) thickness per vertex obtained from 60 training datasets for femur (top row) and tibia (bottom row).

resulting *cartilage thickness maps* are shown in Figure 8.3.

8.2.2   Appearance Cost Functions and Parameter Estimation

Bone.   For bone, we employ the appearance model and continuous cost function as described in Section 3.2.1, with condition $\nabla_{n_v} I(v + l) > g^{bone} > 0$. Image individual parameters t_1^{bone}, t_2^{bone} and g^{bone} are determined automatically via histogram-analysis: The bone intensity threshold t_1^{bone} is computed by analyzing the histogram of intensities inside the initialized SSM (see Fig. 8.4). It is set such that 5% of the voxels that contribute to the histogram have higher intensities. The upper threshold t_2^{bone} is set to the maximum intensity that occurs within the initialized SSM. To determine the gradient threshold g^{bone}, a histogram of gradient magnitudes is computed from the entire image. The gradient threshold is set such that 15% of image voxels exhibit larger gradient magnitudes.

In addition we check for a special case to exclude large, bright artifacts from the intensity window (see Fig. 8.6, left): A weighted sum of 5 Gaussians is fitted to the histogram of intensities by applying the Expectation Maximization (EM) algorithm on a Gaussian mixture model (cf. Sec. 3.2.2). We determine the Gaussian gauss_a with highest peak, i.e. $a := \text{argmaxi}_i \{w_i/\sigma_i\}$, and then we look for the Gaussian gauss_b with highest peak for which holds $\mu_a + \sigma_a < \mu_b - \sigma_b$. If there is such a gauss_b and a value x_0 where gauss_a and gauss_b have equal height, i.e. $w_a \cdot \text{gauss}_a(x_0) = w_b \cdot \text{gauss}_b(x_0)$, and furthermore $\mu_a < x_0 < \mu_b$ and $x_0 < t_1^{bone}$ holds, then t_1^{bone} is set to x_0 and t_2^{bone} is set to $x_0 + 0.25\sigma_b$.

Cartilage.   We employ a cost function that we specifically designed for cartilage in MRI: Costs are high at locations that lie below the learned minimum and above the learned maximum cartilage thickness per vertex in surface normal direction,

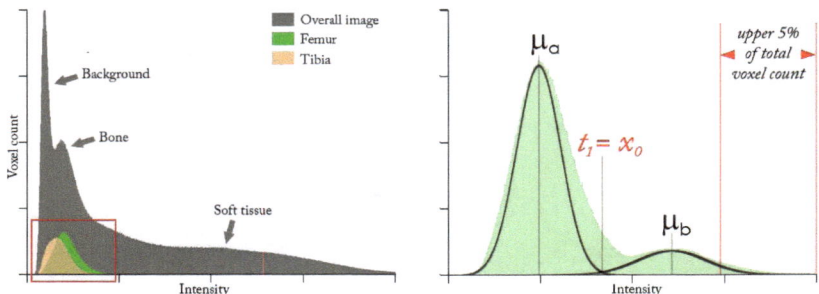

Figure 8.4 *Left:* Schematic histogram of intensities in entire MRI image (gray) and intensities inside initialized SSMs (green: femur, orange: tibia). *Right:* Exemplary histogram of intensities inside an initialized SSM. Parameter estimation: The intensity window $\left[t_1{}^{bone}, t_2{}^{bone}\right]$ is set such that it captures the brightest 5% (voxel count) of intensities. Special case: Of 5 Gaussians fitted via EM, let $gauss_a$ be one with highest peak. We look for $gauss_b$ with brighter mean, i.e. $\mu_b > \mu_a$, non-overlapping standard deviation, and highest peak. If such a $gauss_b$ exists and intersects with $gauss_a$, we set $t_1{}^{bone}$ to the point of intersection, x_0.

$r_{min}(v)$ and $r_{max}(v)$, respectively. Locations within the minimum-maximum range are equipped with costs as in the standard continuous cost function as described in Section 3.2.1, with condition $\nabla_{n_v} I(v+l) < -g^{cart} < 0$. The resulting cost function reads

$$
\phi(v,l) = \begin{cases}
\dfrac{-g^{cart}}{\nabla I_{n(v)}(v+l)} & : I(v+l) \in \left[t_1{}^{cart}, t_2{}^{cart}\right] \text{ and} \\
& \quad \nabla_{n_v} I(v+l) < -g^{cart} \text{ and} \\
& \quad r_{min}(v) < |l| < r_{max}(v) \\[2ex]
2 + \dfrac{|l|}{r_L} & : |l| < r_{min}(v) \vee r_{max}(v) > |l| \\[2ex]
1 + \dfrac{|l|}{r_L} & : \text{else}
\end{cases}
\tag{8.1}
$$

The intensity interval $\left[t_1{}^{cart}, t_2{}^{cart}\right]$ is determined by histogram analysis. The respective histogram is computed from voxels in-between the bone surface and a second surface "bone + maximum cartilage thickness along normals". A weighted sum of five Gaussians is fitted, and the mean ± standard deviation of the Gaussian with highest peak serves as the respective intensity interval. The gradient threshold g^{cart} is set to $10\,\text{mm}^{-1}$ (determined heuristically).

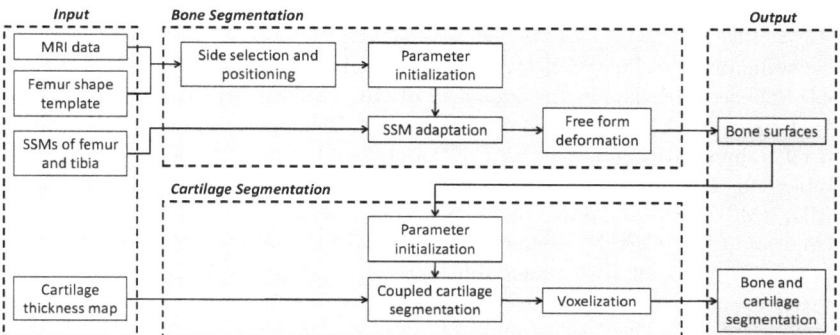

Figure 8.5 Segmentation pipeline for knee bones and cartilage in MRI. See text for explanations.

8.2.3 Multi-object Segmentation Pipeline

The general outline of our automatic segmentation system is shown in Fig. 8.5. It consists of two major parts: (1) An SSM-based pipeline for single bone surfaces as presented for the pelvis (Sec 7.2) applied to femur and tibia. (2) A method for simultaneous segmentation of adjacent structures via shared displacement directions (cf. Sec. 3.5), applied to tibial and femoral cartilage. In the following, we describe the modifications with which we adapted these two methods to the application-specific situation of bone and cartilage in knee MRI.

To detect the correct side of the body, i.e. left or right knee, we perform GHT twice, with a shape template of the left and right distal femur, respectively. The left template mirrored at the mid-sagittal plane yields the right template. The transformation with best match after both GHT runs yields the side of the body and the initial transformation T_0 to position the SSMs of the bones in the data. Given the transformation T_0 of an SSM, the parameters of the bone appearance model are initialized as described in Section 8.2.2. SSM deformation followed by GraphCuts is performed independently for femoral and tibial bone: In contrast to CT (cf. Sec. 8.1), there are significantly less "misleading" image features at the bone-marrow-interface of the adjacent bone, which makes single object segmentation feasible. For cartilage segmentation, the surfaces of femur and tibia are coupled with shared displacement directions and deformed with multi-object GraphCuts employing the cartilage cost function described in Section 8.2.2.

8.2.4   Results and Discussion

For evaluation, 40 clinical MRIs acquired before knee replacement were made available as benchmark data by the organizers of MICCAI 2010 workshop *Medical Image Analysis for the Clinic – A Grand Challenge* (cf. Heimann et al. (2010)) in addition to 60 training MRIs (cf. Sec. 8.2.1). Detailed evaluation results are presented in Table 8.2. The average symmetric mean and root mean square surface distances (MD, RMSD) were computed for the *union of bone and the respective cartilage* for femur and tibia. Cartilage segmentation accuracy is quantified by Jaccard overlap error (JOE) and relative volume difference (RVD) assessed in specific *regions of interest*, namely areas on the respective bone that are in mechanical contact with an adjacent bone. See (Heimann et al., 2010) for details. Benchmark evaluation employs a scoring system as introduced in Section 6.1, featuring the five specific error measures described above. The average score of our auto-segmentation system for knee bones and associated cartilage on 40 benchmark MRIs is 69.2. In the respective on-site segmentation contest at MICCAI 2010, we achieved a comparable score of 68.7 on 10 knee MRIs that were not made available beforehand. This suggests that no notable over-fitting was caused by manual tuning of parameters (cf. Sec. 6.5).

Figure 8.7 shows exemplary segmentation results. Bone segmentation errors are mostly due to relatively large mismatches of the SSM at the proximal and distal end of the MRI data. In these regions image features are weak or missing due to intensity inhomogeneities that are characteristic for MRI (see Fig. 8.6b,d). Furthermore, mismatches occur at large osteophytes (see Fig. 8.7b). Cartilage segmentation errors, if not due to errors of the respective bone segmentation, can be attributed to very low contrast that appears between adjacent cartilages in some datasets (see Fig. 8.7c and 8.6c,d).

In summary, in this section we have proposed a fully automatic pipeline for segmentation of knee bones and cartilage from MRI. The pipeline employs statistical shape models of bones, an application-specific shape and appearance model of cartilage, and shape-constrained free deformations of coupled meshes with multi-object GraphCuts. On benchmark image data, our approach competes well with related work in terms of segmentation accuracy. As for future work, intensity inhomogeneities in MRI data have to be dealt with. Furthermore, a crucial question is how to better model pathological shape changes like osteophytes.

8.3   Conclusion

In this chapter, we have proposed fully automatic segmentation pipelines for articulated joints in CT and MRI. Pipelines are assembled from methods described in Chapter 3.

No.	Femur bone MD [mm]	RMSD [mm]	Scr	Tibia bone MD [mm]	RMSD [mm]	Scr	Femur cartilage JOE [%]	RVD [%]	Scr	Tibia cartilage JOE [%]	RVD [%]	Scr	Total Scr
1	0.53	0.79	77.3	0.52	0.81	66.6	18.5	-4.6	80.7	22.8	-13.1	63.9	72.1
2	0.46	0.67	80.6	0.47	0.65	71.5	26.4	-17.9	46.1	18.4	-12.1	67.9	66.5
3	0.63	0.90	73.7	0.42	0.65	73.1	25.1	14.2	55.3	19.9	-4.0	83.3	71.4
4	0.53	0.76	77.7	0.42	0.59	74.3	32.9	0.3	83.8	20.9	-11.0	69.0	76.2
5	0.48	0.73	79.3	0.39	0.59	75.2	20.8	16.9	51.1	22.9	-12.6	64.9	67.6
6	0.69	1.07	69.9	0.41	0.67	72.8	33.4	36.5	34.3	27.5	-10.3	67.6	61.1
7	0.49	0.79	78.0	0.52	0.75	67.7	21.7	-3.1	82.6	18.5	13.3	65.5	73.5
8	0.75	1.27	65.7	0.46	0.70	70.8	21.9	-13.3	58.9	26.7	-18.0	52.6	62.0
9	0.59	0.89	74.7	0.44	0.69	71.4	27.4	-13.7	55.3	27.0	-15.0	58.3	65.0
10	0.59	0.95	73.7	0.43	0.68	72.2	19.4	-13.2	60.4	30.2	-28.9	36.7	60.7
11	0.53	0.84	76.6	0.44	0.76	70.1	19.5	-2.6	84.7	29.1	-7.4	72.5	76.0
12	0.41	0.68	81.4	0.36	0.52	77.7	17.7	3.6	83.5	22.4	16.5	57.4	75.0
13	0.75	1.21	66.7	0.58	0.91	62.4	36.3	12.5	54.1	36.2	44.4	34.1	54.3
14	0.58	0.89	74.8	0.45	0.70	71.2	26.1	9.3	66.3	14.9	7.0	79.6	73.0
15	0.48	0.70	79.8	0.46	0.74	69.8	20.5	5.4	77.8	25.4	7.2	74.5	75.5
16	0.47	0.74	79.4	0.35	0.51	78.2	20.2	8.4	71.1	23.3	-3.0	83.7	78.1
17	0.54	0.89	75.7	0.41	0.61	74.0	27.9	21.0	38.2	24.8	-16.8	55.8	60.9
18	0.50	0.75	78.6	0.37	0.56	76.6	22.5	4.0	80.1	24.6	1.3	86.7	80.5
19	0.53	0.83	76.8	0.53	0.86	65.0	22.8	14.9	54.8	25.7	-14.4	60.1	64.2
20	0.66	0.92	72.7	0.47	0.72	69.9	26.7	7.2	70.9	38.4	-35.0	33.1	61.6
21	0.45	0.71	80.1	0.36	0.59	76.1	24.3	4.6	77.9	18.2	-4.9	82.3	79.1
22	0.51	0.82	77.4	0.37	0.60	75.7	25.2	6.1	73.9	13.2	-2.7	88.9	79.0
23	0.58	0.85	75.5	0.62	0.97	59.9	28.8	17.8	45.3	36.6	3.8	76.4	64.3
24	0.49	0.75	78.8	0.43	0.68	72.1	24.0	1.5	85.3	16.3	-1.4	90.1	81.6
25	0.59	0.87	74.9	0.60	1.05	58.8	20.7	0.6	88.8	20.0	5.7	79.8	75.6
26	0.61	0.92	73.8	0.42	0.68	72.5	18.3	-3.5	83.4	23.0	14.4	61.3	72.8
27	0.59	0.84	75.4	0.50	0.73	68.8	36.4	-8.9	62.3	31.9	-14.6	57.1	65.9
28	0.68	1.15	69.0	0.43	0.70	71.6	27.4	3.9	78.1	16.3	-6.9	79.1	74.5
29	0.53	0.83	76.8	0.43	0.67	72.3	17.6	-1.1	89.2	26.5	10.4	67.7	76.5
30	0.65	0.96	72.3	0.45	0.73	70.3	27.1	13.5	55.9	38.1	-22.2	39.1	59.4
31	0.62	0.94	73.4	0.43	0.66	72.3	32.4	7.7	67.0	27.9	0.8	86.1	74.7
32	0.48	0.71	79.6	0.44	0.63	72.7	31.9	7.0	68.8	30.0	18.9	49.3	67.6
33	0.47	0.74	79.4	0.43	0.69	72.1	20.9	-9.2	68.8	25.0	0.5	88.0	77.1
34	0.65	1.27	68.0	0.35	0.53	77.5	39.4	56.8	31.5	17.4	9.5	73.5	62.6
35	0.57	0.92	74.6	0.61	1.06	58.2	25.6	7.8	69.9	29.3	-25.2	37.2	60.0
36	0.51	0.76	78.2	0.38	0.58	75.6	15.2	-2.7	86.7	21.8	-12.0	66.6	76.8
37	0.79	1.43	62.7	0.58	0.99	60.8	25.2	6.5	73.1	30.8	17.5	51.7	62.1
38	0.48	0.73	79.3	0.36	0.58	76.3	17.6	4.4	81.7	19.4	8.6	74.3	77.9
39	0.64	1.01	72.0	0.97	1.79	31.7	37.5	4.0	73.2	45.3	-15.4	49.5	56.6
40	0.47	0.71	79.9	0.44	0.67	71.8	18.1	1.3	88.4	24.8	16.3	56.8	74.2
41	0.65	0.97	72.3	0.49	0.77	68.2	23.7	-14.5	55.2	20.8	-8.8	73.3	67.3
42	0.53	0.76	77.8	0.55	0.77	66.4	33.3	33.3	34.3	41.4	9.6	62.7	60.3
43	0.56	0.85	75.8	0.51	0.76	67.8	28.4	2.0	81.9	19.9	-9.3	72.9	74.6
44	0.71	1.19	67.9	0.63	1.01	58.9	32.4	-7.8	66.6	39.2	-38.0	32.8	56.5
45	0.46	0.69	80.4	0.34	0.52	78.3	26.5	6.5	72.6	23.5	17.6	54.7	71.5
46	0.53	0.82	76.9	0.46	0.76	69.4	33.6	4.2	74.5	27.3	-15.8	56.7	69.4
47	1.08	1.36	57.5	0.73	0.98	56.2	62.2	9.3	49.2	33.0	-0.9	83.8	61.7
48	0.55	0.80	76.8	0.49	0.78	68.2	23.3	7.0	72.9	27.8	-15.2	57.6	68.9
49	0.47	0.73	79.5	0.39	0.55	76.0	34.4	33.5	33.8	27.3	17.7	52.9	60.6
50	0.52	0.79	77.6	0.33	0.51	79.0	19.5	13.5	59.6	17.3	-1.3	89.8	76.5
Avg	0.57	0.88	75.1	0.47	0.73	69.7	26.4	5.8	66.8	25.8	-3.1	65.2	69.2

Table 8.2 Error metrics and scores (Scr) for 40 test cases. Last row: Average errors and scores.

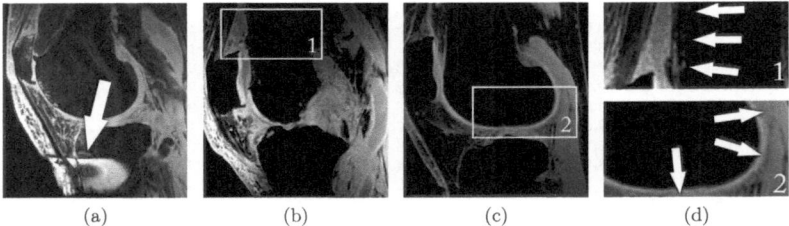

(a)   (b)   (c)   (d)

Figure 8.6 Problem cases: (a) Large bright artifact introducing "false" image features. (b) Intensity inhomogeneities typical for MRI. (c) Similar intensities for cartilage and neighboring soft tissues. (d) Zoom on details framed in (b) and (c). Arrows point to barely visible tissue interfaces.

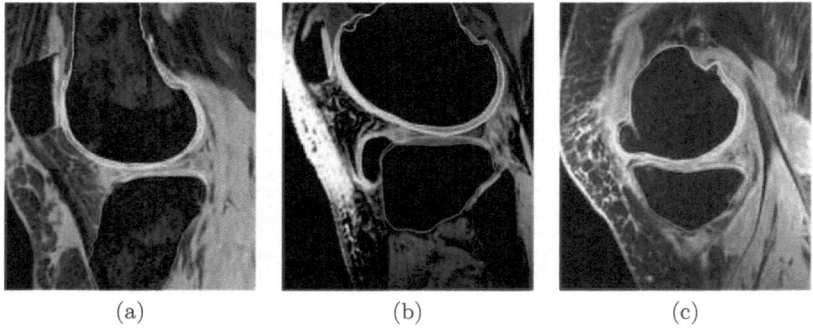

(a)   (b)   (c)

Figure 8.7 Exemplary results: (a) Good result for bone and cartilage. (b) Good cartilage result, but large pathologic protuberance (osteophyte) missed in tibia result. (c) Good bone result, but cartilage too thick for tibia / too thin for femur in joint gap due to low contrast.

A focus is on *GraphCuts* as presented in Sections 3.4.2 and 3.5: We have shown that *multi-object* GraphCuts improve accuracy of hip joint segmentations in CT as compared to *single-object* GraphCuts. This particularly applies to the femoral head, for which ambiguous image features appear inside the adjacent hip bone.

Furthermore, we have proposed a pipeline for fully automatic segmentation of knee bones and cartilage from MRI. Apart from exploiting the benefits of multi-object GraphCuts, the knee pipeline comes up with an application-specific model of cartilage thickness, as well as application-specific algorithms for automatic estimation of MRI appearance parameters.

Hip and knee joint segmentations compete well with related work in terms of accuracy as far as a direct comparison is possible (cf. Chapter 6). However, further improvements can be achieved concerning the hip joint: Chapter 9 reveals that the acetabular rim is affected by the *visibility problem* as described in Section 4.1, and more accurate segmentations can be obtained with *fastODDS* as described in Section 4.3.

Chapter 9

ODDS for Segmentation of Highly Curved Structures

Contents

In Chapter 4 we propose *ODDS*, a method that can handle free deformations of highly curved structures by allowing omnidirectional displacements for all vertices of a surface mesh during mesh deformation. In this chapter, we present a quantitative evaluation on clinical data showing that ODDS outperform traditional mesh adaptation along line segments (e.g. surface normals) in terms of segmentation

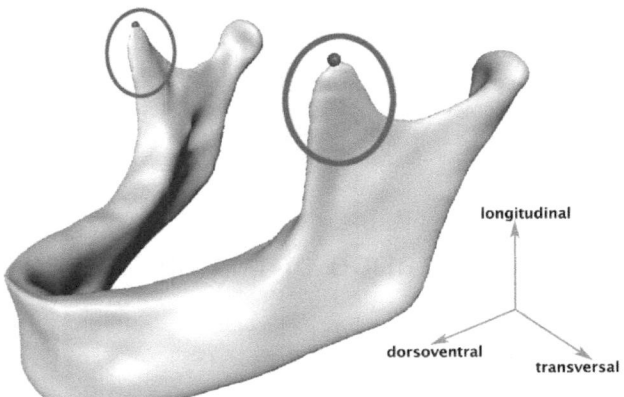

Figure 9.1 Exemplary mandibular bone anatomy and respective body axes. Red/gray circles: Left and right coronoid process. Red/gray dots: Respective tip points.

accuracy for tip- and ridge-shaped structures.

To save run-time and memory as required by ODDS, we developed a hybrid approach, *fastODDS*, also presented in Chapter 4. This approach employs omni-directional displacements adaptively, i.e. only where high curvature calls for them, and traditional unidirectional displacements elsewhere. In this chapter we show in an evaluation on clinical data that fastODDS achieve the same segmentation accuracy as ODDS in regions of high curvature, while requiring only half the run-time and memory.

An additional benefit of fastODDS is that they allow for simultaneous adaptation of multiple, adjacent meshes, i.e. multi-object segmentation. We present an evaluation on clinical data showing that fastODDS outperform traditional multi-object mesh adaptation along line segments.

9.1 Experimental Setup

To evaluate ODDS on clinical data, we apply it to 106 CBCTs of the mandibular bone. The mandibular coronoid process serves as an exemplary tip-shaped structure. See Fig. 9.1 for an exemplary mandibular bone anatomy. To evaluate fastODDS, we apply it to two cohorts of clinical image data: (1) The above 106 CBCTs of the mandibular bone to assess the differences to ODDS, and (2) 49 CTs

	$\#V$	$\bar{e}$	$2r$	δ_S	$\#S$	$\#\tilde{S}$	$[t_1, t_2]$	g	n_f	$\mathfrak{g}$
Mandible	8561	1.2	15	0.4	41272	2188	$[350, 800]$	75	6	6
Hip Bone	14008	2.1	20	0.5	48078	1714	$[120, 720]$	25	10	10

Table 9.1 Application specific parameters are the number of vertices $\#V$ of the deformable mesh (which determines the average edge length of mesh triangles $\bar{e}$ [mm]), the diameter $2r$ [mm] and sampling distance δ_S [mm] of sets of displacements, the number of sample points $\#S$ and MRF labels $\#\tilde{S}$ resulting for ODDS/fastODDS, the intensity window $[t_1, t_2]$ [HU] of the cost function, the gradient magnitude threshold g [1/mm] and filter length n_f (in number of edges) and geodesic distance $\mathfrak{g}$ [mm] for definition of the OmniD region.

of the hip bones – with the acetabular rim as an exemplary ridge-shaped structure in a multi-bone environment – to assess the multi-object ability of fastODDS.

On the mandibular CBCTs, we perform a comparative evaluation not only between ODDS and fastODDS, but also to results generated with FreeBand and GraphCuts (cf. Sec. 3.4), as well as repeated, i.e. *iterative* GraphCuts (*iGraphCuts*). Note that performing GraphCuts iteratively loosens the respective shape constraints. On the hip bone CTs, we compare fastODDS to coupled multi-object GraphCuts (cf. Sec. 8.1).

If not specified otherwise, we use the same experimental setup as for experiments on synthetic data as described in Sec. 4.2.5. Intensity parameters t_1, t_2 and g employed for computing costs $\phi(v, s) \in \mathbf{R}$ are set per application (see Tab. 9.1). Parameters for iGraphCuts are set as for the respective GraphCuts experiment. Adaptations with iGraphCuts were performed iteratively, with 30 steps. Whenever we employ fastODDS, we detect ridges automatically as described in Appendix 4.3.5, with significance 0.04 mm^{-1} and curvature threshold 0.1 mm^{-1}. Furthermore, fastODDS parameters for omnidirectional displacements are set as in the respective ODDS experiments, while fastODDS parameters for unidirectional displacements are set as in the respective GraphCuts experiment. Table 9.1 lists the values of application specific parameters.

A prerequisite for reproducible evaluations of segmentation results is an automatic identification of locations of interest on gold standard and automatically generated surfaces. We present respective methods for the mandibular coronoid process and acetabular rim of the hip bones in Sections 9.1.1 and 9.1.2.

9.1.1 Identification of the Mandibular Coronoid Process

For gold standard as well as automatically determined mandible surfaces, we extract the right coronoid processes as the region of the mesh that below above 1/2 of the extension of the mandible in transversal direction, between 1/3 and 2/3 of extension in dorsoventral direction, and above 2/3 in longitudinal direction (cf. Fig. 9.1). Extraction of the left coronoid process worked analogously. We identified the tip point as the up-most vertex in longitudinal direction.

9.1.2 Acetabular Rim Delineation on Surface Meshes

For gold standard as well as automatically determined hip bone surfaces, we extract the acetabular rim automatically as explained in the following. The statistical shape model of the hip bones we employ for initial segmentation contains a particular *patch* that defines the acetabulum, cf. Sec. 7.2. This acetabular patch is inherent on every initial segmentation, and is preserved during deformation with any of the adaptation methods we employ in this work. The boundary of the deformed acetabular patch serves as an initial estimate of the acetabular rim. It is represented by a set of vertices that are connected by edges which form a closed contour. Starting from this initial estimate, automatic detection of the acetabular rim proceeds as described in Algorithm 9.

Algorithm 9 Automatic delineation of the acetabular rim on pelvis surfaces.

1: Define an approximate "rim-plane" via plane-fit to the initial acetabular rim estimate.

2: For each vertex on the initial rim estimate, sample a set of points on the hip bone surface in direction perpendicular to the rim within some geodesic distance.

3: Define a *cost* per sample point as the signed distance from the approximate rim-plane in *outward* direction. The outward direction of the acetabular rim plane can be determined by means of the orientation of the acetabular patch.

4: Construct a graph: For every pair of neighboring vertices, connect corresponding sample points by and edge in the graph; Connect sample points $\pm$ the corresponding one to achieve the desired amount of regularization;

5: Perform Dijkstra's algorithm (Dijkstra, 1959) to obtain the minimum-cost rim. The result serves as automatically detected acetabular rim.

We evaluated automatic rim delineation vs. manually defined landmarks on 147 surface meshes stemming from manual and automatic hip bone segmentations. Resulting error measures are listed in Tab. 9.2. Results indicate that our algorithm captures the course of the acetabular rim with an average root mean square curve distance of about the average edge length of the respective surface meshes (cf. Table 9.1). Hence we consider it suitable for use in evaluation (cf. Sec. 9.2.2), where it ensures reproducibility of results.

rms [mm]	hd [mm]	%>1mm [%]
1.21(0.31)	**3.06**(0.90)	**34.84**(12.37)

Table 9.2 Automatic acetabular rim delineation: Average root mean square (rms) and Hausdorff (hd) distance from manually defined landmarks, and percentage of distance above 1 mm (%> 1mm), assessed on 147 hip bone surfaces. Standard deviations in brackets.

9.2 Results

9.2.1 Mandibular Coronoid Process

In a quantitative evaluation on 106 mandible CBCTs with voxel size $0.3 \times 0.3 \times 0.3 \, \text{mm}^3$ (cf. Sec. 7.3.1) we compared ODDS, fastODDS, FreeBand, iGraphCuts and GraphCuts results to gold standard surfaces obtained from manual segmentations. Initial meshes were generated automatically via SSM deformation, as described in Section 7.3.

For all omnidirectional displacements, we gave slight preference to displacements that point further "outward" in curvature gradient direction $\nabla k_1(v)$, where k_1 is the magnitude of the maximum principal curvature of the deformable surface (cf. Hildebandt et al. (2005)). Note that $\nabla k_1(v)$ is perpendicular to the surface normal at v. We achieve this preference by adding to $\phi(v, s)$ a small cost proportional to $r_S - s \cdot \nabla k_1(v)/\|\nabla k_1(v)\|$. To also exploit information about curvature gradient for all unidirectional displacements, we do not use normal displacements, but rotate displacement directions from surface normals towards curvature gradient direction. We found that a rotation angle of 45 degrees yields the best results for the mandible. By considering the curvature gradient of the mesh, we intend to find "better" points on sharp, tip-like target structures in the sense of *anatomical correspondence*, and hence reduce the amount of mesh distortion necessary for accurate segmentation of the tip.

We identify coronoid processes and their tip points automatically as described in Section 9.1.1. As error measures for the coronoid process, we assess the tip-to-tip distances (tip2tip), tip-to-surface distances (tip2surf), and Hausdorff (HD) surface distances, as well as the percentage of two-sided surface distances above 1.2 mm (%>1.2mm). Evaluation results are shown in Table 9.3 and Figure 9.2. As measurements are not normally distributed, we assess the significance of differences between methods by means of Wilcoxon's signed-rank test (cf. Sec. 6.3). Results are discussed in Section 9.3.

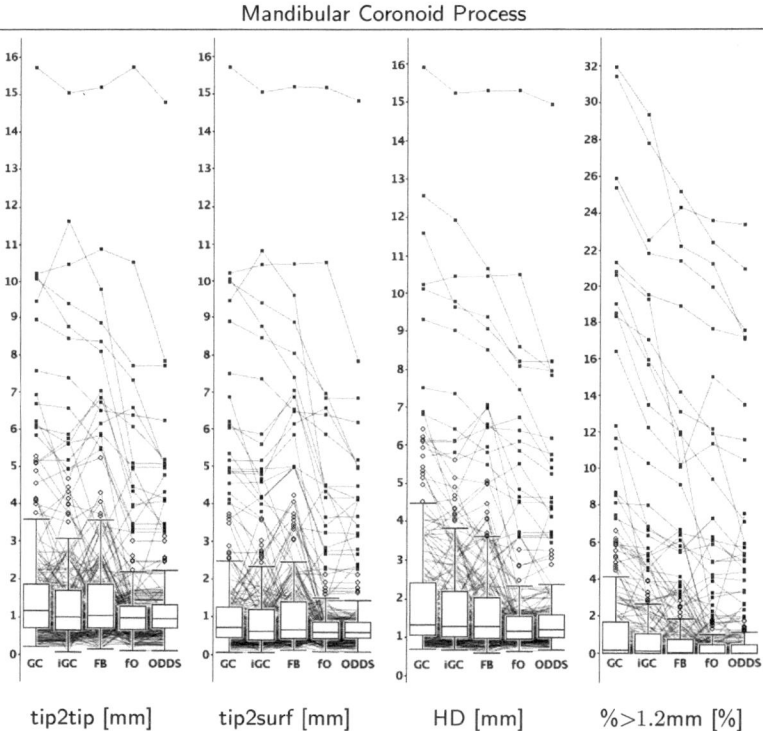

tip2tip [mm]   tip2surf [mm]   HD [mm]   %>1.2mm [%]

Figure 9.2 Box plots of error measures for GraphCuts (GC), iterative GraphCuts (iGC), FreeBand (FB), fastODDS (fO) and ODDS results on coronoid processes as listed in Tab. 9.3. Under-laid parallel coordinate plots draw lines between errors measured for different methods (GC, iGC, FF, fO, ODDS) on corresponding individual cases, e.g. between the tip2tip errors of fastODDS- and ODDS-result on coronoid process no. 189, etc.

	Coronoid Process			
SSM	tip2tip [mm] **2.44**(2.10)	tip2surf [mm] **2.12**(2.15)	HD [mm] **2.76**(2.16)	%>1.2mm [%] **4.48**(7.12)
FreeBand	**1.72**(2.00)	**1.43**(2.02)	**1.98**(1.98)	**1.51**(4.04)
GraphCuts	**1.79**(2.08)	**1.69**(2.17)	**2.22**(2.18)	**2.30**(5.39)
itGraphCuts	**1.66**(2.02)	**1.32**(2.02)	**2.07**(2.01)	**1.83**(4.62)
fastODDS	**1.38**(1.69)	**1.05**(1.66)	**1.70**(1.76)	**1.23**(3.74)
ODDS	**1.35**(1.52)	**1.03**(1.51)	**1.68**(1.59)	**1.17**(3.39)
FreeBand-ODDS	**0.37**	**0.40**	**0.20**	**0.34**
p-value [%]	*<0.01*	*<0.01*	*0.02*	*0.51*
FreeBand-fastODDS	**0.34**	**0.38**	**0.18**	**0.28**
p-value [%]	*0.03*	*<0.01*	*0.04*	*0.20*
itGraphCuts-ODDS	**0.31**	**0.29**	**0.38**	**0.66**
p-value [%]	*<0.01*	*1.27*	*<0.01*	*<0.01*
itGraphCuts-fastODDS	**0.28**	**0.27**	**0.37**	**0.60**
p-value [%]	*0.72*	*3.07*	*<0.01*	*<0.01*
fastODDS-ODDS	**0.03**	**0.02**	**0.02**	**0.06**
p-value [%]	*31.41*	*46.00*	*23.77*	*-43.05*

Table 9.3 Top to bottom: Average error measures (and standard deviation) for initial SSM adaptation (SSM), FreeBand, iterative GraphCuts (itGraphCuts), GraphCuts, fastODDS and ODDS results on 212 coronoid processes and 106 entire mandibles, followed by differences $A - B$ of average error measures for $A, B \in \{$GraphCuts, fastODDS, ODDS$\}$, together with significance levels of difference (p-values) as assessed with Wilcoxon's signed rank test. A positive p-value indicates that B has lower error than A (at the respective level of significance), while a negative sign indicates that A has lower error than B. Significance levels below 5% are highlighted by color/italics.

9.2.2 Acetabular Rim and Hip Bones

In a quantitative evaluation on 49 hip CTs with voxel size $0.9 \times 0.9 \times 1 \, mm^3$ we compared fastODDS and GraphCuts results to gold standard surfaces obtained from manual segmentations.[1] Initial meshes were generated automatically by adaptation of an ASSM of hip bones and femur (Kainmueller et al., 2009a). In case of omnidirectional displacements, we gave slight preference to displacements in surface curvature gradient direction, as described before for the mandible (cf. Sec.9.2.1).

Again, we experimented with unidirectional displacements rotated from surface normals towards curvature gradient directions to also profit from curvature gradient information – however, we found that for the hip bones, this does not improve accuracy. We attribute this to the more complex shape of the hip bones, which

---

[1] Thanks to Markus Heller (Julius Wolff Institute, Charité - Universitätsmedizin Berlin, Germany) for providing image data of the pelvis. Thanks to Jana Malinowski (Zuse Institute Berlin, Germany) for manually segmenting the hip bones.

	Acetabular Rim		Hip Bone	
	HD [mm]	%>1.5mm [%]	HD [mm]	%>1.5mm [%]
ASSM	**5.95**(2.53)	**66.88**(17.64)	**8.44**(2.53)	**26.74**(7.80)
GraphCuts	**5.00**(2.53)	**36.61**(15.98)	**7.07**(2.36)	**2.39**(1.79)
fastODDS	**4.69**(2.75)	**22.91**(15.34)	**6.92**(2.40)	**1.88**(1.65)
GC-fO	**0.32**	**13.70**	**0.15**	**0.52**
p-value [%]	*1.04*	*<0.01*	5.58	*<0.01*

Table 9.4 Top: Average error measures (and standard deviation) for initial ASSM adaptation as well as GraphCuts and fastODDS results on 98 acetabular rims and 98 hip bones. Bottom: Average differences of GraphCuts (GC) and fastODDS (fO) errors and respective levels of significance (p-values) as assessed with Wilcoxon's signed rank test. Significance levels below 5% are highlighted by color/italics.

exhibit concave and convex structures in close proximity and do not show such a sharp and long tip as the mandibular coronoid process. Consequently, we stick to normal unidirectional displacements.

For a reproducible delineation of the acetabular rim, we compute it *automatically* as described in 9.1.2 on both gold standard segmentations and mesh adaptation results. As error measures for the acetabular rim, we assessed the Hausdorff *curve distance* (HD) as well as the percentage of distance above 1.5 mm (%>1.5mm). Furthermore, we assessed the Hausdorff *surface distance* (HD) as well as the %>1.5mm measure for the whole hip bones. Evaluation results for both acetabular rim and whole hip bone are shown in Table 9.4. As with the mandible, error measures are not normally distributed, and hence we performed Wilcoxon's signed-rank test to assess significant differences. Additionally, Figure 9.3 shows the averaged directional *distance difference maps* for visualizing location-specific differences in accuracy between GraphCuts and fastODDS.

9.2.3 Run-time and Memory Requirements

All experiments were performed on a single 3GHz core with 8GB main memory. Table 9.5 lists the average performance of all methods applied to clinical data. MRF optimization (Komodakis et al., 2008) took between one and six seconds in all ODDS- and fastODDS experiments. Computation of the cost function ϕ was more time-consuming, accounting for more than 90% of the run-time of ODDS and fastODDS as stated in Table 9.5. FastODDS and iterative GraphCuts have comparable run-time.

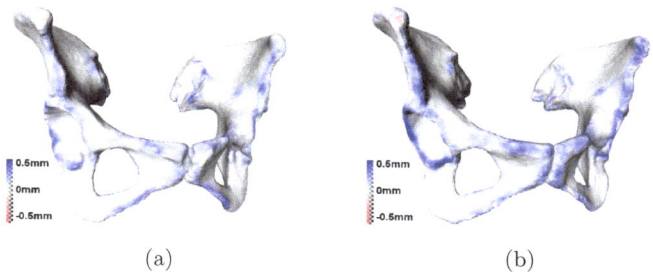

(a)   (b)

Figure 9.3 FastODDS on hip bones, compared to GraphCuts: *Differences* of directional surface distances (GraphCuts-fastODDS) from/to gold standard, averaged over 49 cases. (a) Difference of distances from gold-standard to results. (b) Difference of distances from results to gold-standard. On average, fastODDS perform better than GraphCuts in blue regions, while GraphCuts perform better than fastODDS in red regions.

[sec / GB]	ODDS	fO	GC	iGC	FB
Mandible	149 / 4.6	85 / 2.2	3 / 0.9	90 / 0.9	6 / 0.4
Hip Bone	-	319 / 5.4	18 / 2.3	-	-

Table 9.5 Performance (computation time in seconds / maximum memory requirement in GB) for ODDS, fastODDS (fO), GraphCuts (GC), iterative GraphCuts (itGC) and FreeBand (FB) averaged for 106 mandibles and 98 hip bones.

9.3 Discussion

9.3.1 Segmentation Accuracy

Mandible.   Experiments on CBCTs of the mandible show that both ODDS and fastODDS are able to produce more accurate segmentations of tip-like structures as compared to conventional deformable mesh approaches. On 212 mandibular coronoid processes, ODDS and fastODDS significantly outperform the GraphCuts and FreeBand approach (cf. Table 9.3). Here, normal displacements often exhibit the visibility problem. Figure 9.4 shows exemplary results.

A comparison of ODDS and fastODDS on the mandibular coronoid processes reveals no statistically significant difference for any error measure (cf. Table 9.3, last row). However, the parallel-coordinate plots that underlay the box plots in Figure 9.2 show that there are some individual cases with considerable differences between ODDS and fastODDS error measures. We conclude that fastODDS is not guaranteed to produce equally accurate results in the individual case, but overall

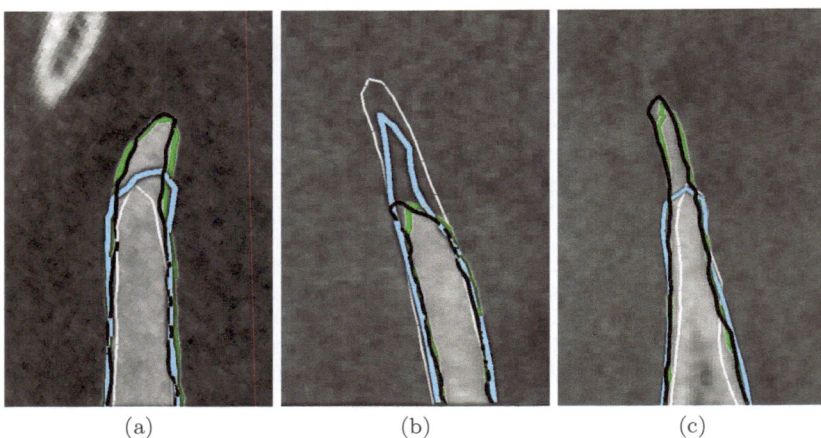

(a)   (b)   (c)

Figure 9.4 ODDS vs. FreeBand: Exemplary results on coronoid processes of the mandible. Contours: Black: Gold standard. White: Initial mesh. Green/gray: ODDS result. Blue/light gray: FreeBand result. While the visibility problem causes inaccurate mesh deformations with FreeBand, ODDS overcomes this limitation, yielding more accurate segmentations.

does not perform significantly different than ODDS.

As for the whole mandible surface, we found an evaluation of error measures to be "overshadowed" by regions on the initial segmentation that are too far away from the target structure in the sense that it does not lie within either ball-shaped or linear sets of candidate locations. These are regions that exhibit misleading or missing image features – the best one can do is to keep them at their initial position, i.e. not deform them at all after initial SSM-based segmentation. This holds for the region around the teeth, where the teeth themselves are potentially mistaken as features, and also the chin, which often lies outside the field of view of the CBCT scanner and hence exhibits no features at all. Table 9.6 states evaluation results on 106 mandibles in terms of Hausdorff errors. While average Hausdorff errors suggest a slight tendency in favor of ODDS/fastODDS, differences between methods are not statistically significant for *any* couple of methods, including initial SSM-based segmentation.

Hip bones.   Experiments on CTs of the pelvis show that fastODDS are able to produce very accurate segmentation of ridge-like structures in a multi-object environment. On 98 acetabular rims of the hip bones, multi-object fastODDS significantly outperform the multi-object GraphCuts approach (cf. Table 9.4). Here,

SSM	FB	GC	iGC	fO	ODDS
7.36	**7.30**	**7.36**	**7.33**	**7.14**	**7.14**
(3.03)	(2.99)	(3.01)	(3.00)	(2.71)	(2.69)

Table 9.6 Average Hausdorff error in mm, assessed on 106 mandible surfaces for initial SSM-based segmentation, FreeBand (FB), GraphCuts (GC), iterative GraphCuts (iGC), fastODDS (fO) and ODDS.

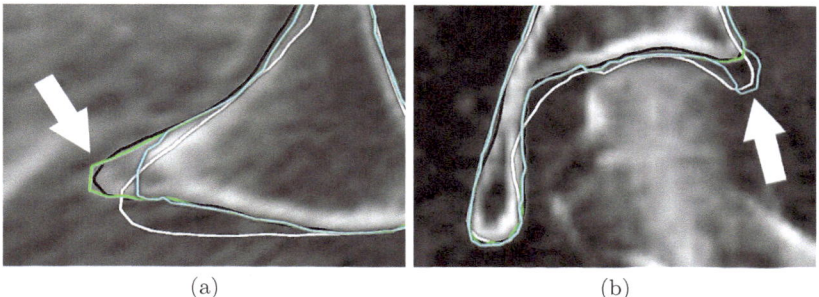

(a)   (b)

Figure 9.5 Exemplary results of fastODDS on acetabular rim of the pelvis. Contours: Black: Gold standard. White: Initial mesh. Green: FastODDS result. Blue: GraphCuts result. While fastODDS produce accurate segmentations, GraphCuts do not reach the respective image features that are located (a) in outward direction and (b) in inward direction from the initial mesh.

unidirectional displacements often struggle with restricted visibility, as shown in Figure 9.5.

FastODDS also perform better than GraphCuts in terms of error measures evaluated on the whole hip bones. As for the Hausdorff error measure, the relatively small improvement from initial (SSM-based) to resulting (GraphCuts and fastODDS) segmentations suggests that the p-value stemming from the comparison of GraphCuts and fastODDS, namely 5.58%, may, as for the mandible but less prominent, be influenced by regions on the initial segmentation that are too far away from the target structure to be within reach of either ball-shaped or linear search range. This observation is also reported in Section 7.2.

9.3.2 Comparability of Regularization

For GraphCuts, differences of displacement lengths on neighboring vertices are "for free" up to the shape constraint parameter c, while larger differences are impossible (see Sec. 3.4.2). For ODDS, differences of displacements on neighboring vertices are "for free" or cost the minimum non-zero distance penalty up to a Euclidean norm of $\delta_{\tilde{S}}$, while the penalty increases cubically for larger differences (see Sec. 4.2.3). FreeBand regularizes the surface mesh and not displacements themselves. However, the maximum edge length that can occur on a needle-shaped (i.e. infinitesimally thin tip) mesh is bounded by displacement stepsize and internal smoothing weight.

To achieve comparable regularization, we set the regularization parameter c of GraphCuts to the displacement block sampling distance $\delta_{\tilde{S}}$ as set in the respective ODDS/fastODDS experiment. Furthermore, FreeBand is parametrized such that when stretching a needle-like tip, the maximum achievable edge length is double the average initial edge length $\bar{e}$, and hence the maximum difference of displacements is $\bar{e}$. For all comparisons involving FreeBand, i.e. for all experiments on mandible data, $\delta_{\tilde{S}} = \bar{e}$, and hence in summary $\delta\tilde{S} = c = \bar{e}$ (cf. Table 9.1).

We think this allows for a fair comparison of methods. However, to make sure that the superior accuracy of ODDS/fastODDS is not an effect of "more or less" regularization, we performed GraphCuts not only with $c = \delta_{\tilde{S}}$, but with c ranging from the sampling distance δ_S up to an absurdly large $10\delta_S = 5\,\text{mm}$ in nine extra experiments on the hip bones. Considering segmentation accuracy, significant improvements of fastODDS over GraphCuts as stated via colored entries in Tab. 9.4 hold for *any* of the respective GraphCuts results.

Cutting the sampling distance by half for unidirectional displacements, i.e. $\delta_L = 0.5\delta_S$ (cf. Sec. 4.2.5), was intended to compensate for a potential advantage of omnidirectional displacements in terms of an effective denser sampling in surface-normal direction due to additional adjacent sampling points. It *did* slightly improve error measures for GraphCuts results – however, the accuracy of ODDS/fastODDS could not be reached, not even with still smaller (nor bigger) sampling distances from 0.25 to $1\delta_S$.

9.3.3 Influence of Mesh Resolution

The flexibility of the deformable mesh is determined by regularization, namely the "tolerated distance" δ_{tol} between neighboring displacements as set via regularization parameters (cf. Sec. 9.3.2), and mesh resolution, namely the average edge length $\bar{e}$ of mesh triangles. Tolerated distance, divided by triangle edge length, $\delta_{tol}/\bar{e}$, serves as a measure for mesh flexibility. However, coarser mesh resolution at constant mesh flexibility in terms of $\delta_{tol}/\bar{e}$ has a smoothing effect on the displacement field. To this end, to assess the influence of mesh resolution at constant flexibility $\delta_{tol}/\bar{e} = 1$ as set in our original experiments (see Tab. 9.1), we performed an additional evaluation of fastODDS, iterative GraphCuts and FreeBand on a series of different mesh resolutions. We achieve different resolutions with an approach for isotropic re-meshing described by Surazhsky and Gotsman (2003). The resulting Hausdorff distances on 212 mandibular coronoid processes are plotted in Fig. 9.6. For fas-

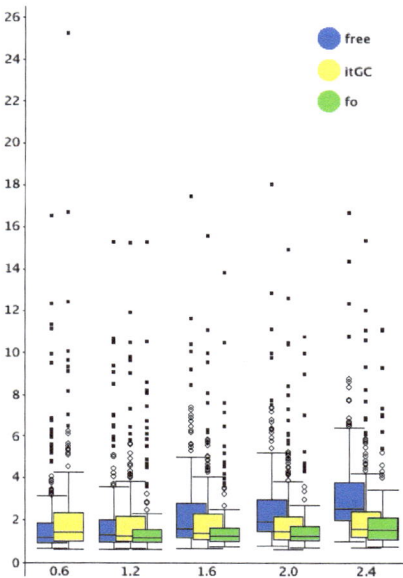

Figure 9.6 Comparison of fastODDS (fo), iGraphCuts (itGC) and FreeBand (free) at a series of mesh resolutions. X-axis: Average edge length [mm] of mesh triangles. Y-axis: Hausdorff distance [mm] assessed for 212 coronoid processes of the mandible. Mandible meshes contain about 34100, the original 8561, 5000, 3100 and 2300 vertices, respectively.

tODDS, we did not evaluate the finest mesh resolution due to unbearable memory requirements ($\geq$ 64 GB). The significance statements given in Table 9.3 hold for *any* resolution on which fastODDS were performed, and furthermore also for the comparison between fastODDS at 1.2 mm edge length and iterative GraphCuts and FreeBand at the smaller edge length of 0.6 mm.

9.3.4   Consistency of Deformed Meshes

Avoiding self-intersections of the deformable mesh is crucial for approaches that employ unidirectional displacements together with *iterative* search for appearance match (FreeBand, iGraphCuts). This is because "loops" in the deformable mesh can invert surface normals and hence render successive displacement directions unfeasible, if not counterproductive. To this end, for iterative approaches, we pre-

	ODDS	fastODDS	GraphCuts
Coronoid Process	**30.8**	**22.2**	**7.6**
	(29.2)	(23.1)	(17.4)
Hip Bone	-	**305.81**	**311.70**
		(140.28)	(163.07)

Table 9.7 Average number of self-intersections in deformed surface mesh, assessed for ODDS, fastODDS and GraphCuts on 212 mandibular coronoid processes, and for fastODDS and GraphCuts on 98 hip bones. Standard deviation in brackets.

vent self-intersections with the method proposed in (Kainmueller et al., 2007), and hence no self-intersections occur (while consequently the deformable mesh may "get stuck"). Instead, for approaches that perform search for appearance match just once (ODDS, fastODDS, GraphCuts), self-intersections of the deformable mesh do not necessarily affect segmentation accuracy. However, they do indicate some sort of quality of mesh deformation. To this end, we assessed the number of self-intersections for the mandibular coronoid process and hip bone results as presented in Sections 9.2.1 and 9.2.2, respectively. Each triangle edge that intersects with a non-adjacent triangle counts as a self-intersection. Table 9.7 lists the results. While fastODDS and ODDS exhibit more self-intersections than GraphCuts on the mandibular coronoid processes, fastODDS produces slightly less self-intersections than GraphCuts on the hip bones.

As discussed in Section 4.2, with parameter values as set for our experiments, omnidirectional sets of candidate displacements of neighboring vertices overlap heavily. Overlap is particularly heavy for the mandible, where the average edge length of meshes equals the "tolerated difference" of displacements on edge-connected vertices, i.e. $\bar{e} = \delta_{\tilde{S}}$ (cf. Table 9.1). Furthermore individual sets of candidate locations most probably contain a whole region (two-dimensional manifold) of the target surface. Therefore we consider the numbers of self-intersections as generated by ODDS and fastODDS unexpectedly little. We conclude that MRF binary potentials serve well their purpose of preventing inconsistent displacements of adjacent vertices.

9.3.5 Run-time and Memory Requirements

A comparison of ODDS and fastODDS on the mandibular bone shows that fastODDS require less than half the memory, while being almost twice as fast as ODDS. The run-time of fastODDS is comparable to iterative GraphCuts. In general, the gain in performance achieved by fastODDS depends on the "curvedness" of the anatomical structure of interest. Hence we hypothesize that the gain is even bigger for structures like the heart or the liver, where a higher percentage of the structure exhibits low curvature, while it may be little to none on highly folded structures like the cerebral cortex or the intestinal mucosa.

9.4 Conclusion

A quantitative evaluation on clinical data of the mandibular bone showed that ODDS significantly outperform traditional mesh adaptation methods in terms of segmentation accuracy in regions with high curvature. We showed that fastODDS achieve the same segmentation accuracy as ODDS, while requiring only half the run-time and memory.

We showed on clinical data of the hip bones that multi-object fastODDS significantly outperform multi-object GraphCuts in terms of segmentation accuracy in the region of the acetabular rim.

To further improve run-time of fastODDS, future work will focus on a more efficient computation of the appearance cost function ϕ for omnidirectional displacements via parallelization and exploitation of overlapping domains.

Chapter 10

Extrapolation and Atlas-based Segmentation of Leg Muscles

Contents

This chapter describes two pipelines that employ deformations of volumetric grids as described in Chapter 5.

Section 10.1 deals with automatic reconstruction of anatomical landmarks adjacent to the pelvic bones from CT, as published in Seim et al. (2009). We compare different approaches for *geometric* reconstruction of landmarks, i.e. image appearance is not modeled for the landmarks. Instead, landmark locations are extrapolated from surface meshes of the pelvic bones. Landmark reconstruction accuracy is assessed and compared in an evaluation on 49 hip CTs. Results show that extrapolation with Mean Value Coordinates (MVCs, cf. Sec. 5.1.4) yields improved reconstruction accuracy as compared to extrapolation with a compound SSM thanks to the ability of MVCs to transfer free deformations of surface meshes to surrounding structures.

Section 10.2 employs extrapolation of mesh deformations to volume deformations

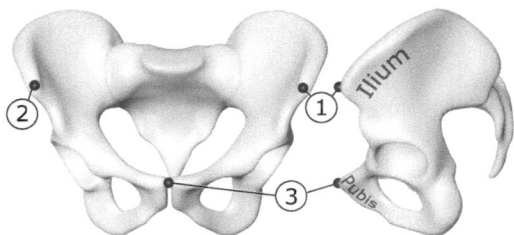

Figure 10.1 Pelvis shape with anatomical landmarks LASIS (1), RASIS (2) and symphysis (3).

within a pipeline for segmentation of structures that are not suitable for SSM modeling, yet are located adjacent to SSM-modellable structures (cf. Chapter 5), and exhibit faint appearance characteristics in the image data: Namely we deal with segmentation of individual leg muscles in CT and exploit their proximity to a compound of articulated bones. Our approach builds upon the work of Berger (2011). Muscle segmentations are estimated via extrapolation of bone deformations, and subsequently refined by means of atlas image registration (cf. Sec. 5.2). To the best of our knowledge we propose the first fully automatic approach for segmentation of individual muscles from CT. Accuracy is assessed in an evaluation on the musculus gluteus medius in 20 hip CTs, each segmented with 10 atlases, making a total of 200 experiments. Each of the total of 30 CTs involved depicts a different individual. Comparative evaluation of affine, polyaffine and MVC extrapolation (cf. Sec. 5.1) of bone surface mesh deformations reveals unexpected deficits of extrapolation methods that take into account bone articulation: We hypothesize that polyaffine and MVC extrapolation do not yield anatomically plausible deformations, and conclude that future work should focus on alternative, *physically motivated* extrapolation methods.

10.1   Extraction of Anatomical Landmarks of the Pelvic Bones

The determination of anatomical landmarks is an essential step in morphological analysis of a solitary bone, and a key prerequisite for defining reference systems to assess relative positions of bones forming a joint. The anterior pelvic plane (APP) is a reference plane defined by three anatomical landmarks, namely the left and right anterior superior iliac spines (LASIS and RASIS) and the pubic symphysis (SYM) as shown in Figure 10.1. An accurate determination of the APP is mandatory for referencing the orientation of the acetabulum (Lewinnek et al., 1978). This is a key measure that enables the orthopedic surgeon to assess important changes in anatomy and resulting biomechanical conditions which are either due to a disease or a consequence of surgery (Lewinnek et al., 1978; Heller et al., 2007).

Accurate automatic detection of anatomical landmarks facilitates the assessment of anatomical changes in individual subjects, and thereby allows for a monitoring of large patient populations. Thus, potentially subtle yet relevant distinctive features may be identified that either predispose to the development of degenerative joint diseases or fuel their further progression (Gregory et al., 2007).

This section evaluates methods for automatic detection of anatomical landmarks in CT data, namely LASIS, RASIS and SYM. The SYM landmark in general does not lie on the bone surface, while the ASIS landmarks do.

Related work has focused on automatic landmark detection methods that exploit appearance characteristics of landmarks in medical image data: With the goal of providing an orthopedic planning tool, Ehrhardt et al. (2004) proposed an atlas image registration approach for detecting pelvic landmarks in CT. Landmark localization based on extremal differential properties of images like ridges, corners or saddles, was introduced by Wörz and Rohr (2006). Izard et al. (2006) suggest an algorithm for landmark detection based on a probabilistic model of image intensities. A method that exploits spatial relationships and appearance characteristics of landmarks learned from training data was introduced by Dikmen et al. (2008).

Related work as cited above performs *image-based* reconstruction of landmarks, i.e. it exploits specific appearance characteristics. In contrast, in this section we describe and evaluate automatic methods for *geometric* reconstruction of pelvic landmarks based on previously reconstructed surfaces of the pelvis. This geometric approach allows for landmark reconstruction in any situation that comes with geometric reconstructions of the pelvis, potentially also in case of *2D-3D reconstruction* (Lamecker et al., 2006a) where image data is not present around landmarks.

Given CT data, a fully automatic framework for pelvis segmentation as proposed in Section 7.2 generates the pelvis reconstructions. With a focus on the accuracy of approaches based on Mean Value Coordinates (MVCs, cf. Sec. 5.1.4), we evaluate and compare a series of geometric methods for landmark reconstruction on 49 clinical CT datasets for which gold standard landmarks were defined manually by multiple observers.

10.1.1 Methods

The methods described in the following exhibit different degrees of generality, ranging from problem-specific to generic.

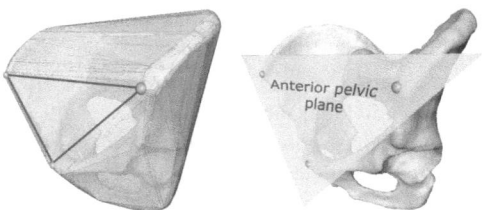

Figure 10.2 Convex hull of the pelvic bones (left) including the APP with minimum distance to both ilia and the symphysis (right).

Convex Hull

The Convex Hull method is based on the assumption that the landmarks of interest are the most prominent anterior points of the pelvic anatomy, defining the anterior pelvic plane (APP). Thus, they are also part of the convex hull of the pelvis. The method first computes the triangulated convex hull H of a reconstructed pelvis surface. The APP is determined by identifying the triangle t of H whose vertices have the smallest sum of distances from pubis, the ilium, and right ilium (see Figure 10.2). These regions define *patches* on pelvis surface meshes. Thus for each of the three patches, the distance to the closest vertex of a triangle of H can be determined. The vertex on the left ilium patch of the reconstructed pelvis surface with minimum distance to the APP is considered the LASIS. In case there are multiple vertices with minimal distance, we simply select the one with smallest index. The RASIS is defined analogously. For the pubis, two nearest points to the APP are determined, one for each hip bone. The midpoint between the two nearest points is considered the symphysis landmark SYM. The convex hull method represents a problem-specific approach that does not require any training landmarks.

Compound SSM

We build a compound SSM of the pelvic bones and landmark locations from training data that we also employed to build an SSM of the pelvic bones alone as described in Section 7.2. The respective pelvic landmarks were manually defined in each training image. With the resulting compound SSM, automatic reconstruction of the pelvic bones from CT as described in Section 7.2 produces surface meshes of the the pelvis with consistent mesh topology as well as SSM-extrapolated landmark positions (cf. Sec. 3.1.5).

Mean Value Coordinates

(a) Mean value weights of landmarks contained in the deformed compound SSM are computed using the respective SSM-deformed pelvis surface mesh as control mesh. Subsequent free deformations of pelvis meshes yield accurate segmentations of the pelvic bones (cf. Sec. 7.2). Free deformations do not alter mesh topology. Hence mean value weights allow for a transfer of landmark coordinates to freely deformed pelvis surface meshes (cf. Sec. 5.1.4).

(b) Mean value weights are computed for each training landmark using the respective training pelvis surface as control mesh. Landmarks are transferred to freely deformed pelvis meshes by means of averaged mean value weights (cf. Sec. 5.1.4). Averaging mean value weights reduces the influence of *outliers* i.e. inaccurately defined landmarks in the training set. This method is applicable also in case training landmarks exist only for a subset of the training set and hence a compound SSM cannot be built from the whole training set. In consequence it is the most generic of all described methods.

10.1.2 Results and Discussion

For evaluation 49 CTs with resolution $0.9 \times 0.9 \times 1 \,\mathrm{mm}^3$ were available (cf. Sec. 9.2.2). For all datasets three experts manually located landmarks, namely left and right ASIS as well as the symphysis.[1] We chose one of these three sets of manually placed landmarks as reference. The other two sets serve for assessing the inter-observer variability of manual landmark placement.

We compare four automatic landmark extraction methods as described in Section 10.1.1, which we refer to as Hull (convex hull method), SSM (compound SSM method), SSM+MVC (compound SSM plus MVCs) and aMVC (average MVCs). As error measure we assess Euclidean distances of automatically reconstructed landmarks to gold standard landmarks. Additionally we assess the angle θ between the APP defined by automatically reconstructed landmarks and the respective gold standard APP. Table 10.1 lists average error measures and standard deviations.

Inter-observer variability for the ASIS landmarks ranges between 2.3 and 3.2 mm in terms of Euclidean distance and between 0.5° and 0.7° in terms of APP angle. As for automatic methods, the Hull method performs best in terms of Euclidean distance as well as APP angle, followed by the aMVC method. The SSM+MVC method performs considerably better than "pure" compound SSM extrapolation, but worse than average MVCs.

The convex hull method produces results that are only slightly less accurate than manual landmark reconstructions performed by a second observer. However,

[1] Thanks to Alexander Wurl, Philippe Moewis and Markus Heller (Julius Wolff Institute, · Charité - Universitätsmedizin Berlin, Germany) for manually locating the pelvic landmarks.

	ASIS	SYM	θ
	mm (std)		deg (std)
User2	**2.3** (1.5)	**3.2** (1.2)	**0.5** (0.4)
User3	**2.7** (2.4)	**2.5** (1.2)	**0.7** (0.6)
Hull	**3.8** (2.5)	**3.6** (1.9)	**1.0** (0.9)
SSM	**7.1** (3.6)	**4.6** (2.0)	**2.5** (1.4)
SSM+MVC	**6.8** (3.6)	**4.3** (2.0)	**1.7** (1.2)
aMVC	**6.6** (3.2)	**3.8** (2.0)	**1.3** (1.0)

Table 10.1 Landmark reconstruction: Evaluation results on 49 CTs. Average inter-observer variability (User2, User3) and average error metrics of automatic methods (as described in Section 10.1.1) in bold font, standard deviations in brackets. Euclidean distances of ASIS and SYM in mm, APP angle differences in deg.

it is also the most problem-specific method and cannot be transferred to landmarks not defined via the convex hull of the pelvis. Approaches based on training landmarks yield less accurate results, particularly for the ASIS landmarks. This may be attributed to training landmarks exclusively stemming from low resolution data (5 mm slice thickness, cf. Sec. 7.2). Furthermore, although reference ASIS landmarks lie on reference pelvis surfaces they were not exploited for establishing surface correspondences.

Part of the landmark position error in all automatic methods can be attributed to few cases in which automatic reconstruction of the pelvis shows inaccuracies in the region of the iliac crest. Furthermore, methods involving pelvis training meshes rely on the quality of point-to-point surface correspondences established for SSM generation. Future work needs to assess the influence of surface correspondences as well as landmark distances to reference surfaces on reconstruction accuracy.

Overall, aMVC produces results that are more accurate than SSM+MVC results, and only slightly less accurate than Hull results. Training landmarks are not required for the whole set of training data. We conclude that averaged Mean Value Coordinates are a valuable generic tool for automatic detection of landmarks which are located in close vicinity of an SSM-modellable structure.

10.2 Segmentation of Leg Muscles

Patient-specific biomechanical simulations as performed e.g. for implant planning require reconstructions of musculoskeletal anatomy from medical image data (Heller et al., 2001). For accurate biomechanical modeling of muscle forces, straight lines connecting attachment sites are not sufficient; There is at least the need for wrapping muscle center lines around obstacles (Audenaert and Audenaert, 2008), and ideally for volumetric models (Blemker and Delp, 2005; Jolivet et al., 2008). Meth-

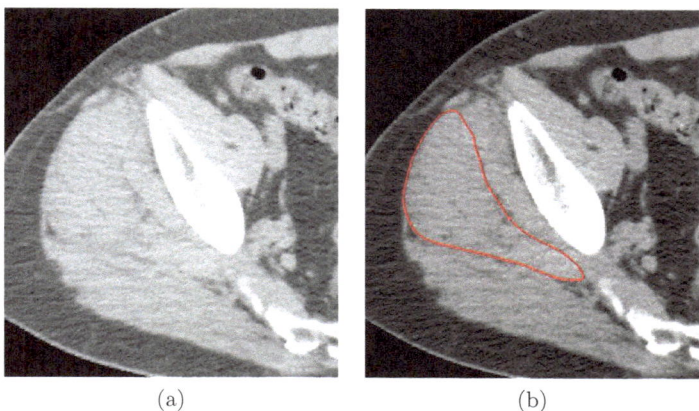

(a)   (b)

Figure 10.3 (a) Gluteus medius in an axial slice of CT. (b) Gold standard segmentation shown as red contour.

ods for automatic reconstruction of patient-specific muscles from image data can contribute to any therapy planning system being used to a greater extent in clinical practice. In orthopedics, CT is indicated for imaging bone fractures, because in MRI, dry material (bone) and air (fractures) may not be distinguishable (see e.g. Bui and Taira (2010)). Exploiting CT to also reconstruct patient specific muscle anatomy would be of great benefit for cost effective surgery and therapy planning.

There is very few work on automatic segmentation of muscles from CT. This is a challenging task as muscles are low-contrasted in CT and hence adjacent muscles are hard to separate visually. Figure 10.3 shows an exemplary slice of hip CT and a gold standard segmentation of the musculus gluteus medius. Jolivet et al. (2008) propose a *semi-automatic* method that can interpolate volumetric segmentations of individual limb muscles from manual segmentations in few axial slices. Chung et al. (2009) propose a method for *automatic* segmentation of muscle tissue in a specific axial cross-section of CT, i.e. in 2d. Apart from working exclusively in 2d, individual muscles are not discriminated by this approach - it deals with *muscle tissue* segmentation. A lot more has been achieved concerning the automatic segmentation of muscles from MRI, which exhibits much better contrast; see e.g. Gilles et al. (2006) and Baudin et al. (2012).

To the best of our knowledge, this section presents the first work on fully automatic 3d segmentation of muscles from CT. In addition to large inter-individual shape variations, muscle shape is exposed to three types of *intra-individual* variation, namely articulation, tension and training. This complex mixture of large, non-linear, intra-individual variations renders muscle shape not straightforwardly suitable for modeling with SSMs (cf. Chapter 5). Instead, we follow a 3D atlas registration approach for segmentation (cf. Sec. 5.2). This way we avoid explicit shape modeling while at the same time coping with the challenge of low contrast between individual muscles in CT.

We are dealing with human lower limb muscles, where *articulated* 3D image registration methods appear suitable to cope with joints. Methods for articulated registration have been proposed, yet for different applications, as e.g. for registration of neck images (du Bois d'Aische et al., 2005), time-series of mouse images (Papademetris et al., 2005), and histological slices (Arsigny et al., 2005). Methods are generally composed of two steps: First, bones and in some cases skin are aligned by means of linear transformations. Second, the resulting transformations are extrapolated onto the surrounding soft tissue. Some methods proceed with a third step that performs intensity-based non-rigid registration of the initially deformed data from step 2. Methods most notably differ in how they propagate transformations in step 2, namely via FEM propagation of piecewise rigid trafos (du Bois d'Aische et al., 2005), or by blending piecewise rotations to achieve a continuous (Papademetris et al., 2005) or even differentiable and invertible (Arsigny et al., 2005) resulting deformation.

Our approach for muscle segmentation follows this general "extrapolation followed by registration" pipeline, as detailed in Section 10.2.1. Additionally, as a very first step, we perform automatic bone segmentation, which renders our whole muscle segmentation pipeline fully automatic.

We demonstrate the potential of our approach in an evaluation on 20 target CTs, where we segment the musculus gluteus medius with 10 reference atlases. Results are described and discussed in Section 10.2.2. Figure 10.4 shows the anatomy of the gluteus medius. It spreads between iliac crest and trochanter major. It is purely surrounded by other muscle tissue towards the distal end (approaching the trochanter major), exhibiting low contrast in CT. On the medial/anterior side it is partially adjacent to the hip bone, and to fat tissue on the lateral/posterior side towards the proximal end, which exhibit better contrast. However, proximally, the gluteus medius is also adjacent to muscles, namely to the gluteus minimus (medial/anterior) and maximus (lateral/posterior).

10.2.1 Segmentation Pipeline

First, we perform fully automatic multi-bone segmentation in atlas- as well as target images as described in Section 8.1. This yields point-to-point correspondences between reference and target bone surfaces – i.e. deformations that map reference- to target bone surfaces.

Second, we extrapolate this deformation to yield a volumetric deformation of the whole atlas image (cf. Sec. 5.1), containing the muscles. As for the extrapolation

Ilium,
posterior
surface of
iliac crest

**Gluteus
medius**

Greater
trochanter

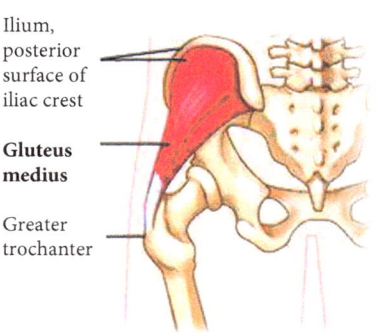
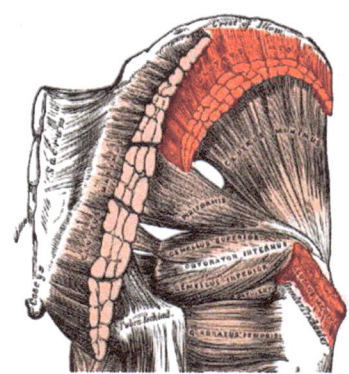

Figure 10.4 Gluteus Medius Anatomy. Left: Mosby's Medical Dictionary, 8th edition. 2009, Elsevier. Right: Attachment sites, Gray's Anatomy of the Human Body, 1918.

technique, we compare affine to Polyaffine Transformations and Mean Value Coordinates as described in Section 5.1. For extrapolation via affine transformation we compute the affine transformation which optimally aligns the compound of all reference bones to the compound of target bones. For Polyaffine Transformations, we consider two patches, one for each bone, namely femur and hip bone. This way we take into account bone articulation. Weight functions required for Polyaffine Transformations are set to the inverse of Euclidean distances of locations in space to individual bones. All three methods compute extrapolations efficiently and produce continuous deformations, where affine and Polyaffine Transformations are differentiable and invertible, and Mean Value Coordinates yield exact deformations which are differentiable almost everywhere.

Third, we perform image based registration of the initially deformed reference image onto the target image by means of image based non-linear registration. As we are dealing with CT, we employ Sum of Squared Differences (SSD) as similarity measure for registration (cf. (5.5)). To make the similarity measure robust against extreme intensities (stemming e.g. from artifacts induced by metal implants, or air inclusions in the intestines) we crop the intensity range of both reference and target image to a window $[-200, 200]$. As an additional measure to gain robustness, in the atlas image, we only consider the area of the atlas muscle grown by a radius of 7 cm. Furthermore, we exploit bone segmentations known a priori for both reference and target by incorporating an additional penalty term into the overall objective, namely the SSD similarity measure applied on bone label images (where different bones have different labels). To cope with large deformations we expect to be required for inter-subject registration of muscles, we employ a hyperelastic deformation model as described in Burger et al. (2013) (cf. (5.4)). As for the weighting of terms, the length term is weighted with $\alpha_1 := 1500$, the volume term with $\alpha_2 := 100$, and the penalty term for bone label images with $\beta := 10000$. We determined these values empirically.

Finally, the resulting overall grid deformation (registration∘extrapolation) is applied to a surface mesh representations of the reference muscle segmentations to obtain a segmentation of the target muscle.

10.2.2 Results and Discussion

We evaluate the segmentation pipeline on the musculus gluteus medius in 20 target CTs of the pelvic area, where we employ 10 reference CTs as atlases, and compare three bone extrapolation methods, namely MVCs, polyaffine (PA), and affine transformations, as described in Sec. 10.2.1. Our image data stems from a post-op study on patients of ages between 46 and 77 that underwent total hip replacement surgery in one hip joint.[2] We perform automatic segmentation of the gluteus medius on the opposite, "healthy" side – however, the hip implants cause severe metal artifacts that do spread onto this side. Resolution is $0.8 \times 0.8 \times 5 \, mm^3$ for all datasets.

As error measures, we assess the dice similarity coefficient (Dice) as well as the absolute value of relative volume difference (ARVD). Error measures are assessed for all three extrapolation methods, initially after extrapolation, and finally after registration. Apart from an analysis of the $10 \cdot 20 = 200$ individual segmentation results, we also evaluate 20 "fused" segmentations obtained by majority voting. Results are listed in Table 10.2 and shown as box plots in Figure 10.5. Besides volumetric error measures, we also asses the mean surface distance error w.r.t. normalized pelvis height – i.e. which error occurs at which height, leading from trochanter major to iliac crest (cf. Figure 10.6).

Image based registration significantly improves segmentation accuracy as compared to pure extrapolation. This also holds in regions where the muscle is surrounded by adjacent muscles, i.e. in regions of low contrast to surrounding tissue, namely the distal end of the gluteus medius (cf. Fig. 10.6). As for "absolute" accuracy, the average Dice coefficient and absolute relative volume difference achieved with affine extrapolation, image based registration and majority voting over the 10 references are 89,3 % and 4,4 %, respectively (cf. Tab. 10.2 bottom right). We expect that more sophisticated label fusion techniques will yield a considerable increase in accuracy compared to simple majority voting as performed in this work.

---

[2] Thanks to Markus Heller (Julius Wolff Institute, Charité - Universitätsmedizin Berlin, Germany) for providing image data of the pelvis. Thanks to Alexander Wurl (Julius Wolff Institute) for manually segmenting the gluteus medius.

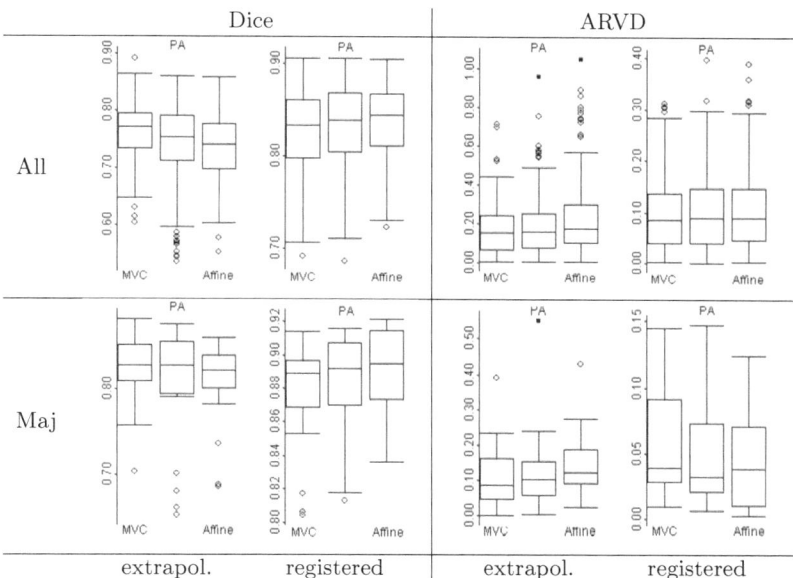

Figure 10.5 Gluteus Medius segmentation results on 20 targets with 10 references. Dice and ARVD error measures for extrapolated and registered results as box plots. Extrapolation with Mean Value Coordinates (MVC), Polyaffine Transformations (PA), and Affine transformation (Affine). Top row (All): 200 individual results. Bottom row (Maj): 20 majority vote results.

	Dice/extrapolated			Dice/registered		
	MVC	PA	Affine	MVC	PA	Affine
all	76.2 (5.1)	74.1 (7.0)	73.3 (5.8)	82.5 (4.4)	83.2 (4.2)	83.6 (4.0)
maj	82.1 (4.3)	80.4 (7.1)	80.9 (5.1)	87.6 (3.3)	88.4 (3.0)	89.3 (2.3)
	ARVD/extrapolated			ARVD/registered		
	MVC	PA	Affine	MVC	PA	Affine
all	17.2 (12.9)	19.3 (15.9)	23.0 (19.0)	9.5 (7.2)	9.9 (7.3)	10.4 (7.7)
maj	11.1 (9.6)	12.7 (12.3)	14.4 (9.6)	5.7 (4.3)	5.3 (4.9)	4.4 (4.0)

Table 10.2 Gluteus medius results: Average and standard deviation (in brackets) of error measures as presented in Figure 10.5.

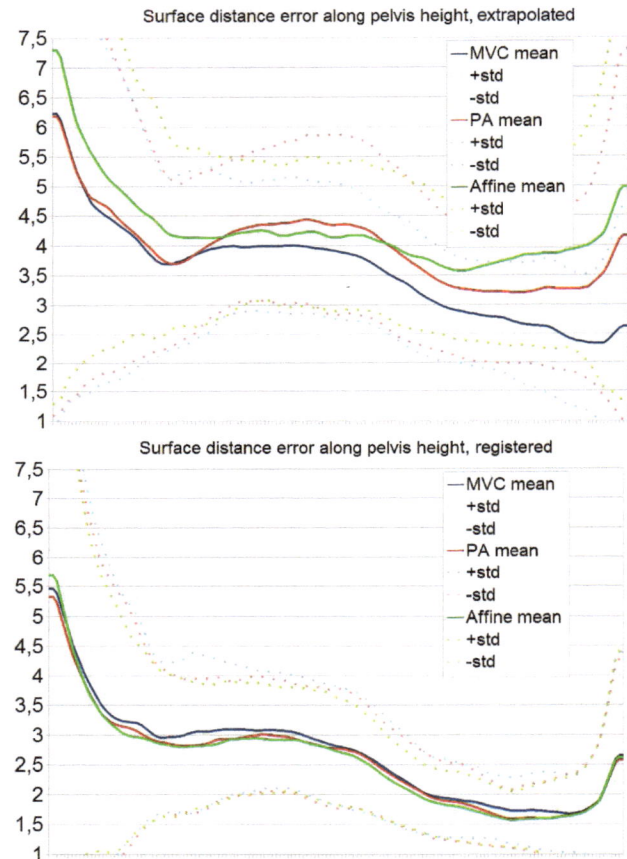

Figure 10.6 Results after extrapolation (left) and registration (right): Average surface distance error plotted over normalized pelvis height. X-Axis: Normalized pelvis height (0 = Distal end of Trochanter Major, 1 = Proximal end of Iliac Crest). Y-Axis: Surface distance error in mm. Comparison of Mean Value Coordinates (MVC), Polyaffine (PA) and affine transform.

As for the particular extrapolation methods, after extrapolation, MVC results are more accurate than PA results, which are more accurate than Affine results. This is what we expected, given the decreasing "amount of bone information" exploited by the respective method. However, surprisingly, things change upside down after registration, where affine extrapolation exceeds PA, which exceeds MVC. We deduce that neither MVC nor PA are able to generate "anatomically plausible" extrapolations. In effect, though MVC and PA themselves are more accurate than affine extrapolation, the hyperelastic deformation model employed in registration seems to not be able to re-capture anatomically plausible shapes, and hence registration ends up less accurate.

One obvious deficiency of both MVC and PA is that muscle attachment sites are not taken into account: If a muscle resides close to a bone, MVC and PA give this bone large influence on the resulting deformed muscle, even if the attachment site – which supposedly has much more influence – is far away. Future work may lead in the direction of better extrapolation of bone deformation by exploitation of attachment sites.

In summary, we consider our work on muscle segmentation preliminary as results suggest that neither of the examined geometric extrapolation techniques yields anatomically plausible deformations. We consider physically motivated deformation models a promising alternative for extrapolation of bone articulation and deformations. This hypothesis is supported by the fact that a hyperelastic deformation model yielded the most accurate muscle segmentation results if *not* preceded by geometric extrapolation methods, although starting from less accurate initial segmentations.

Apart from methodological improvements, to put segmentation accuracy into context, future work has to assess inter-observer variability of muscle segmentations in CT. Comparison to segmentation accuracy reported in related work (on MRI segmentation) is difficult as each work gives different error measures, mostly averaged over a set of different leg muscles. To the best of our knowledge, no individual evaluation of automatic gluteus medius segmentation has been published, neither for CT nor MRI.

10.3 Conclusion

In this chapter we have applied geometric extrapolation techniques to propagate bone surface deformations to surrounding structures. We have proposed a framework for geometric reconstruction of anatomical landmarks of the pelvic bones. Furthermore, we have presented, to the best of our knowledge, the first fully automatic 3d segmentation method for muscles in CT.

Mean Value Coordinates have shown potential as a generic tool for accurate reconstruction of anatomical landmarks which reside in close proximity to SSM-modellable structures. However, neither MVCs nor Polyaffine Transformations appear to be suitable for anatomically plausible extrapolation of bone articulation in joints onto surrounding soft tissue. To this end, future work will assess the potential of alternative, physically motivated models for extrapolation of surface deformations onto volumes.

Conclusions

In this thesis, we have described a toolkit for fully automatic segmentation of anatomical structures in 3d medical image data. Following the Deformable Meshes approach for segmentation, the core tools are statistical shape models and shape-constrained free deformation models.

This thesis impacts the field of medical image processing by contributing novel methods for shape-constrained free mesh deformation as well as application-specific segmentation pipelines which yield significant gains in segmentation accuracy as compared to related work.

We have proposed a novel deformation model for triangle meshes, ODDS. We have shown that ODDS significantly improve segmentation accuracy for tip- or ridge-shaped structures as compared to conventional mesh deformation approaches. Furthermore, we have proposed a mesh coupling algorithm which allows for overlap-free simultaneous deformation of arbitrary adjacent meshes. We have shown that simultaneous deformation of coupled meshes yields improved segmentation accuracy as compared to separate, independent deformation of single meshes.

From our toolkit of methods, we have assembled fully automatic pipelines for segmentation tasks that arise in clinical practice. We have presented thorough quantitative evaluations of segmentation accuracy on clinical image data. Direct comparison of evaluation results to related work requires common pools of benchmark image data and standard error measures. As far as such data is available, our segmentation approach yields high segmentation accuracy: As at June 2014 it ranks first among all fully automatic methods competing in terms of a score on a pool of benchmark liver CTs, and second on a pool of benchmark knee MRIs, respectively.

For two applications, we present, to the best of our knowledge, the first fully automatic 3d segmentation methods with quantitative evaluation, namely for the mandibular nerve channel in CBCT, and for individual leg muscles in CT.

In this thesis we have focused on improving the accuracy of fully automatic segmentation. As for the impact of this thesis on clinical practice, future work needs to assess and evaluate time savings regarding manual labor on segmentations.

We have applied our Deformable Meshes toolkit for segmentation of medical images. However, our methodological contributions are generic and thus not limited to medical applications. Tomographic images depicting structures with distinguished shape, together with a need for their segmentation, appear in numerous other fields of application, e.g. in biology, biochemistry and paleontology. In addition, the field of computer vision comes with a range of respective 2D segmentation tasks. We are convinced that any application which conforms to our basic assumptions on shape characteristics will benefit from the novel methodologies we proposed in this thesis.

Publications

This thesis is based on the following publications as first or equally contributing second author:

D. Kainmueller, H. Lamecker, B. Weber, M. Heller, H.-C. Hege, and S. Zachow. Omnidirectional Displacements for Deformable Surfaces. *Medical Image Analysis*, 17(4):429 – 441, 2013.

D. Kainmueller, H. Lamecker, H. Seim, S. Zachow, and H.-C. Hege. Improving Deformable Surface Meshes through Omni-directional Displacements and MRFs. In *MICCAI*, volume 6361 of *LNCS*, pages 227 – 234. Springer, 2010.

H. Seim, D. Kainmueller, H. Lamecker, M. Bindernagel, J. Malinowski, and S. Zachow. Model-based Auto-segmentation of Knee Bones and Cartilage in MRI Data. In *Medical Image Analysis for the Clinic: A Grand Challenge*, pages 215 – 223, 2010.

D. Kainmueller, H. Lamecker, S. Zachow, and H.-C. Hege. An Articulated Statistical Shape Model for Accurate Hip Joint Segmentation. In *EMBC*, pages 6345–6351, 2009.

D. Kainmueller, H. Lamecker, and S. Zachow. Multi-object Segmentation with Coupled Deformable Models. *Annals of the British Machine Vision Association*, 5: 1–10, 2009.

H. Seim, D. Kainmueller, M. Heller, S. Zachow, and H.-C. Hege. Automatic Extraction of Anatomical Landmarks from Medical Image Data: An Evaluation of Different Methods. In *ISBI*, pages 538–541, 2009.

D. Kainmueller, H. Lamecker, H. Seim, M. Zinser, and S. Zachow. Automatic Extraction of Mandibular Nerve and Bone from Cone-Beam CT Data. In *MICCAI*, volume 5762 of *LNCS*, pages 76–83. Springer, 2009.

H. Seim, D. Kainmueller, M. Heller, H. Lamecker, S. Zachow, and H.-C. Hege. Automatic Segmentation of the Pelvic Bones from CT Data Based on a Statistical Shape Model. In *VCBM*, pages 93–100, 2008.

D. Kainmueller, T. Lange, and H. Lamecker. Shape Constrained Automatic Segmentation of the Liver based on a Heuristic Intensity Model. In *Medical Image Analysis for the Clinic: A Grand Challenge*, pages 109–116, 2007.

Bibliography

M. Alexa. Linear Combination of Transformations. *ACM Transactions on Graphics*, 21(3):380–387, 2002. (Cited on page 71.)

A. Amini, T. Weymouth, and R. Jain. Using Dynamic Programming for Solving Variational Problems in Vision. *IEEE Transactions on Pattern Analysis and Machine Intelligence*, 12(9):855 –867, 1990. (Cited on page 42.)

V. Arsigny, X. Pennec, and N. Ayache. Polyrigid and Polyaffine Transformations: A Novel Geometrical Tool to Deal with Non-rigid Deformations – Application to the Registration of Histological Slices. *Medical Image Analysis*, 9(6):507–523, 2005. (Cited on pages 70, 71, 72 and 154.)

A. Audenaert and E. Audenaert. Global Optimization Method for Combined Spherical-cylindrical Wrapping in Musculoskeletal Upper Limb Modelling. *Computer Methods and Programs in Biomedicine*, 92(1):8–19, 2008. (Cited on page 152.)

D. H. Ballard. Generalizing the Hough Transform to Detect Arbitrary Shapes. *Pattern Recognition*, 13(2):111–122, 1981. (Cited on page 32.)

P.-Y. Baudin, N. Azzabou, P. Carlier, and N. Paragios. Prior Knowledge, Random Walks and Human Skeletal Muscle Segmentation. In *MICCAI*, volume 7510 of *LNCS*, pages 569–576. Springer, 2012. (Cited on page 153.)

G. Behiels, F. Maes, D. Vandermeulen, and P. Suetens. Evaluation of Image Features and Search Strategies for Segmentation of Bone Structures in Radiographs using Active Shape Models. *Medical Image Analysis*, 6(1):47 – 62, 2002. (Cited on page 42.)

R. Beichel, S. Mitchell, E. Sorantin, F. Leberl, A. Goshtasby, and M. Sonka. Shape- and Appearance-based Segmentation of Volumetric Medical Images. In *International Conference on Image Processing*, volume 2, pages 589–592, 2001. (Cited on page 4.)

J. Berger. Nichtlineare Registrierung zur Segmentierung des Musculus Gluteus Medius in CT Daten. Master's thesis, Universität zu Lübeck, Institute of Mathematics and Image Computing, 2011. Supervised by Dagmar Kainmüller, Hans Lamecker, Nils Papenberg, and Bernd Fischer. (Cited on page 148.)

M. Bindernagel, D. Kainmueller, H. Seim, H. Lamecker, S. Zachow, and H.-C. Hege. An Articulated Statistical Shape Model of the Human Knee. In H. Handels and et al., editors, *Bildverarbeitung für die Medizin*, Informatik aktuell, pages 59 – 63. Springer, 2011. (Cited on pages 68 and 119.)

M. Bindernagel. Articulated Statistical Shape Models. Master's thesis, Zuse-Institut Berlin and Humboldt-Universität zu Berlin, Department of Computer Science, 2013. Supervised by Dagmar Kainmüller, Hans Lamecker, Stefan Zachow, and Beate Meffert. (Cited on pages 68 and 119.)

W. Birkfellner. *Applied Medical Image Processing: A Basic Course*. Taylor & Francis, 2011. (Cited on page 5.)

Å. Björck. *Numerical Methods for Least Squares Problems*. Society for Industrial and Applied Mathematics, 1996. (Cited on pages 33 and 70.)

A. Blake, P. Kohli, and C. Rother. *Markov Random Fields for Vision and Image Processing*. MIT Press, 2011. (Cited on page 54.)

S. S. Blemker and S. L. Delp. Three-Dimensional Representation of Complex Muscle Architectures and Geometries. *Annals of Biomedical Engineering*, 33(5):661–673, 2005. (Cited on page 152.)

M. Bucki, C. Lobos, and Y. Payan. A Fast and Robust Patient Specific Finite Element Mesh Registration Technique: Application to 60 Clinical Cases. *Medical Image Analysis*, 14(3):303 – 317, 2010. (Cited on page 53.)

A. Bui and R. Taira. *Medical Imaging Informatics*. Springer, 2010. (Cited on pages 118, 121 and 153.)

M. Burger, J. Modersitzki, and L. Ruthotto. A Hyperelastic Regularization Energy for Image Registration. *SIAM Journal on Scientific Computing*, 35(1):B132–B148, 2013. (Cited on pages 74, 75 and 156.)

F. Cazals, F. Chazal, and T. Lewiner. Molecular Shape Analysis based upon the Morse-Smale Complex and the Connolly Function. In *Proceedings of the nineteenth Annual Symposium on Computational Geometry*, pages 351–360. ACM, 2003. (Cited on page 65.)

J. Chambers. *Graphical Methods for Data Analysis*. Chapman & Hall statistics series. Wadsworth International Group, 1983. (Cited on page 84.)

G. Chintalapani, L. M. Ellingsen, O. Sadowsky, J. L. Prince, and R. H. Taylor. Statistical atlases of bone anatomy: Construction, iterative improvement and validation. In N. Ayache, S. Ourselin, and A. J. Maeder, editors, *MICCAI*, volume 10 of *LNCS*, pages 499–506. Springer, 2007. (Cited on page 101.)

H. Chung, D. Cobzas, L. Birdsell, J. Lieffers, and V. Baracos. Automated Segmentation of Muscle and Adipose Tissue on CT Images for Human Body Composition Analysis. In *Medical Imaging: Image Processing*, SPIE Conference Series, pages 72610K–72610K–8. SPIE, 2009. (Cited on page 153.)

J. Conway, N. Sloane, and E. Bannai. *Sphere Packings, Lattices, and Groups*. A Series of Comprehensive Studies in Mathematics. Springer, 1999. (Cited on page 54.)

T. Cootes, A. Hill, C. Taylor, and J. Haslam. Use of Active Shape Models for Locating Structures in Medical Images. *Image and Vision Computing*, 12:355–366, 1994. (Cited on page 91.)

T. Cootes and C. Taylor. Statistical Models of Appearance for Computer Vision. Technical report, University of Manchester, Wolfson Image Analysis Unit, Imaging Science and Biomedical Engineering, Manchester M13 9PT, United Kingdom, 2004. (Cited on page 29.)

T. F. Cootes, C. J. Taylor, D. H. Cooper, and J. Graham. Active Shape Models - Their Training and Application. *Computer Vision and Image Understanding*, 61 (1):38–59, 1995. (Cited on pages 7, 23, 26, 28 and 31.)

T. F. Cootes, G. J. Edwards, and C. J. Taylor. Active Appearance Models. *IEEE Transactions on Pattern Analysis and Machine Intelligence*, 23(6):681–685, 2001. (Cited on page 11.)

M. J. Costa, H. Delingette, S. Novellas, and N. Ayache. Automatic Segmentation of Bladder and Prostate Using Coupled 3D Deformable Models. In N. Ayache, S. Ourselin, and A. J. Maeder, editors, *MICCAI*, volume 4791 of *Lecture Notes in Computer Science*, pages 252–260. Springer, 2007. (Cited on page 43.)

H. Coxeter. *Introduction to Geometry*. Wiley, 1969. (Cited on page 20.)

E. W. Dijkstra. A note on two problems in connexion with graphs. *Numerische Mathematik*, 1:269–271, 1959. (Cited on pages 42, 90, 112 and 134.)

M. Dikmen, Y. Zhan, and X. S. Zhou. Joint Detection and Localization of Multiple Anatomical Landmarks Through Learning. In *Medical Imaging: Image Processing*, volume 6915 of *SPIE Conference Series*, 2008. (Cited on page 149.)

A. du Bois d'Aische, M. De Craene, B. Macq, and S. Warfield. An Improved
Articulated Registration Method for Neck Images. In *International Conference
of the IEEE Engineering in Medicine and Biology Society*, volume 7, pages 7668–
71, 2005. (Cited on page 154.)

A. Duysak, J. J. Zhang, and V. Ilankovan. Efficient Modelling and Simulation
of Soft Tissue Deformation using Mass-spring Systems. In *Computer Assisted
Radiology and Surgery*, volume 1256 of *International Congress Series*, pages 337
– 342. Elsevier, 2003. (Cited on page 70.)

D. Eggert, A. Lorusso, and R. Fisher. Estimating 3-D Rigid Body Transformations:
A Comparison of Four Major Algorithms. *Machine Vision and Applications*, 9:
272–290, 1997. (Cited on page 31.)

J. Ehrhardt, H. Handels, W. Plötz, and S. J. Pöppl. Atlas-based Recognition
of Anatomical Structures and Landmarks and the Automatic Computation of
Orthopedic Parameters. *Methods of Information in Medicine*, 43(4):391–397,
2004. (Cited on pages 101 and 149.)

A. X. Falcão, J. K. Udupa, S. Samarasekera, S. Sharma, B. E. Hirsch, and
R. de A. Lotufo. User-Steered Image Segmentation Paradigms: Live Wire and
Live Lane. *Graphical Models and Image Processing*, 60(4):233 – 260, 1998. (Cited
on page 42.)

G. Fischer. *Lineare Algebra*. Vieweg, 2005. (Cited on page 28.)

M. S. Floater, G. Kós, and M. Reimers. Mean Value Coordinates in 3D. *Computer
Aided Geometric Design*, 22(7):623–631, 2005. (Cited on page 72.)

J. Folkesson, E. Dam, O. F. Olsen, P. Pettersen, and C. Christiansen. Automatic
Segmentation of the Articular Cartilage in Knee MRI using a Hierarchical Multi-
class Classification Scheme. In *MICCAI*, volume 8, pages 327–334, 2005. (Cited
on page 122.)

R. Forman. Morse Theory for Cell Complexes. *Advances in Mathematics*, 134(1):
90–145, 1998. (Cited on page 65.)

J. Fripp, S. Crozier, S. K. Warfield, and S. Ourselin. Automatic Segmentation of the
Bone and Extraction of the Bone-cartilage Interface from Magnetic Resonance
Images of the Knee. *Physics in Medicine and Biology*, 52(6):1617–1631, 2007.
(Cited on page 122.)

S. Gerber, T. Tasdizen, and R. Whitaker. Dimensionality Reduction and Principal
Surfaces via Kernel Map Manifolds. In *International Conference on Computer
Vision*, pages 529–536, 2009. (Cited on page 67.)

B. Gilles, L. Moccozet, and N. Magnenat-Thalmann. Anatomical Modelling of the Musculoskeletal System from MRI. In *MICCAI*, volume 4190 of *LNCS*, pages 289–296. Springer, 2006. (Cited on pages 43 and 153.)

B. Glocker, N. Komodakis, N. Paragios, C. Glaser, G. Tziritas, and N. Navab. Primal/Dual Linear Programming and Statistical Atlases for Cartilage Segmentation. In N. Ayache, S. Ourselin, and A. Maeder, editors, *MICCAI*, volume 4792 of *Lecture Notes in Computer Science*, pages 536–543, Berlin, Heidelberg, 2007. Springer Berlin Heidelberg. (Cited on page 122.)

B. Glocker, N. Komodakis, G. Tziritas, N. Navab, and N. Paragios. Dense Image Registration through MRFs and Efficient Linear Programming. *Medical Image Analysis*, 12(6):731 – 741, 2008. (Cited on page 54.)

R. Gonzalez. *Digital Image Processing*. Pearson Education, 3 edition, 2009. (Cited on page 5.)

C. Goodall. Procrustes Methods in the Statistical Analysis of Shape. *Journal of the Royal Statistical Society. Series B (Methodological)*, 53(2):pp. 285–339, 1991. (Cited on page 30.)

J. Gower. Generalized Procrustes Analysis. *Psychometrika*, 40:33–51, 1975. 10.1007/BF02291478. (Cited on page 30.)

H. Graichen, R. von Eisenhart-Rothe, T. Vogl, K.-H. Englmeier, and F. Eckstein. Quantitative Assessment of Cartilage Status in Osteoarthritis by Quantitative Magnetic Resonance Imaging: Technical Validation for use in Analysis of Cartilage Volume and Further Morphologic Parameters. *Arthritis & Rheumatism*, 50 (3):811–6, 2004. (Cited on page 121.)

J. S. Gregory, J. H. Waarsing, J. Day, H. A. Pols, M. Reijman, J. Weinans, and R. M. Aspden. Early Identification of Radiographic Osteoarthritis of the Hip using an Active Shape Model to Quantify Changes in Bone Morphometric Features: Can Hip Shape Tell us Anything about the Progression of Osteoarthritis? *Arthritis & Rheumatism*, 56:3634–3643, 2007. (Cited on page 149.)

B. Haas, T. Coradi, M. Scholz, P. Kunz, M. Huber, U. Oppitz, L. André, V. Lengkeek, D. Huyskens, A. van Esch, and R. Reddick. Automatic Segmentation of Thoracic and Pelvic CT Images for Radiotherapy Planning using Implicit Anatomic Knowledge and Organ-specific Segmentation Strategies. *Physics in Medicine and Biology*, 53(6):1751–1771, 2008. (Cited on page 101.)

G. Hamarneh, J. Yang, C. McIntosh, and M. Langille. 3D Live-wire-based Semiautomatic Segmentation of Medical Images. In J. M. Fitzpatrick and J. M. Reinhardt, editors, *SPIE Conference Series*, volume 5747, pages 1597–1603, 2005. (Cited on page 11.)

H. Handels. *Medizinische Bildverarbeitung.* Vieweg + Teubner, 2nd edition, 2009. (Cited on pages 2, 3, 6, 18, 73, 74 and 82.)

N. Hanssen, Z. Burgielski, T. Jansen, M. Lievin, L. Ritter, B. von Rymon-Lipinski, and E. Keeve. Nerves – Level Sets for Interactive 3D Segmentation of Nerve Channels. In *International Symposium on Biomedical Imaging*, pages 201–204, 2004. (Cited on page 108.)

H.-C. Hege, M. SeeBaß, D. Stalling, and Malte Zöckler. A Generalized Marching Cubes Algorithm Based on Non-Binary Classifications. Technical report, Zuse Institute Berlin, 1997. (Cited on page 21.)

T. Heimann. Optimierung des Segmentierungsvorgangs und Evaluation der Ergebnisse in der medizinischen Bildverarbeitung. Technical Report 145, DKFZ Heidelberg, 2003. (Cited on page 19.)

T. Heimann, I. Wolf, and H.-P. Meinzer. Active Shape Models for a Fully Automated 3D Segmentation of the Liver - An Evaluation on Clinical Data. In R. Larsen, M. Nielsen, and J. Sporring, editors, *MICCAI*, volume 4191 of *LNCS*, pages 41–48. Springer-Verlag, 2006. (Cited on page 91.)

T. Heimann, S. Münzing, H.-P. Meinzer, and I. Wolf. A Shape-Guided Deformable Model with Evolutionary Algorithm Initialization for 3D Soft Tissue Segmentation. In N. Karssemeijer and B. Lelieveldt, editors, *Information Processing in Medical Imaging*, volume 4584 of *LNCS*, pages 1–12. Springer, 2007. (Cited on page 91.)

T. Heimann, B. van Ginneken, M. Styner, Y. Arzhaeva, V. Aurich, C. Bauer, A. Beck, C. Becker, R. Beichel, G. Bekes, F. Bello, G. Binnig, H. Bischof, A. Bornik, P. Cashman, Y. Chi, A. Cordova, B. Dawant, M. Fidrich, J. Furst, D. Furukawa, L. Grenacher, J. Hornegger, D. Kainmueller, R. Kitney, H. Kobatake, H. Lamecker, T. Lange, J. Lee, B. Lennon, R. Li, S. Li, H.-P. Meinzer, G. Nemeth, D. Raicu, A.-M. Rau, E. van Rikxoort, M. Rousson, L. Rusko, K. Saddi, G. Schmidt, D. Seghers, A. Shimizu, P. Slagmolen, E. Sorantin, G. Soza, R. Susomboon, J. Waite, A. Wimmer, and I. Wolf. Comparison and Evaluation of Methods for Liver Segmentation from CT datasets. *IEEE Transactions on Medical Imaging*, 28(8):1251–1265, 2009. (Cited on pages 4, 90, 91 and 98.)

T. Heimann and H.-P. Meinzer. Statistical Shape Models for 3D Medical Image Segmentation: A Review. *Medical Image Analysis*, 13(4):543 – 563, 2009. (Cited on pages 4, 7, 26, 27, 28, 29, 30, 31, 36, 38 and 90.)

T. Heimann, B. J. Morrison, M. A. Styner, M. Niethammer, and S. K. Warfield. Segmentation of Knee Images: A Grand Challenge. In B. van Ginneken, K. Murphy, T. Heimann, V. Pekar, and X. Deng, editors, *MICCAI*, pages 207–214, Bejing, China, 2010. (Cited on pages 85, 118, 122 and 126.)

T. Heimann and H. Delingette. Model-Based Segmentation. In T. M. Deserno, editor, *Biomedical Image Processing*, Biological and Medical Physics, Biomedical Engineering, pages 279–303. Springer Berlin Heidelberg, 2011. (Cited on pages 35 and 36.)

M. Heller, G. Bergmann, G. Deuretzbacher, L. Drselen, M. Pohl, L. Claes, N. Haas, and G. Duda. Musculo-skeletal loading conditions at the hip during walking and stair climbing. *Journal of Biomechanics*, 34:883–93, 2001. (Cited on pages 118 and 152.)

M. O. Heller, J. H. Schröder, G. Matziolis, A. Sharenkov, W. R. Taylor, C. Perka, and G. N. Duda. Musculoskeletal Load Analysis. A Biomechanical Explanation for Clinical Results – and More? *Der Orthopäde*, 36:188–194, 2007. (Cited on page 149.)

K. Hildebandt, K. Polthier, and M. Wardetzky. Smooth Feature Lines on Surface Meshes. In M. Desbrun and H. Pottmann, editors, *Proc. Eurographics Symposium on Geometry Processing*, pages 85–90. Eurographics Association, 2005. (Cited on pages 21, 52, 65 and 135.)

M. Hollander and D. Wolfe. *Nonparametric Statistical Methods*. Wiley Series in Probability and Statistics. Wiley, 1999. (Cited on page 85.)

C. Izard, B. Jedynak, and C. E. L. Stark. Spline-Based Probabilistic Model for Anatomical Landmark Detection. In R. Larsen, M. Nielsen, and J. Sporring, editors, *MICCAI*, volume 4190 of *LNCS*, pages 849–856. Springer, 2006. (Cited on page 149.)

E. Jolivet, E. Daguet, V. Pomero, D. Bonneau, J. D. Laredo, and W. Skalli. Volumic Patient-specific Reconstruction of Muscular System based on a Reduced Dataset of Medical Images. *Computer methods in biomechanics and biomedical engineering*, 11(3):281–90, 2008. (Cited on pages 152 and 153.)

I. Jolliffe. *Principal Component Analysis*. John Wiley & Sons, Ltd, 2005. (Cited on page 28.)

T. Ju, S. Schaefer, and J. Warren. Mean Value Coordinates for Closed Triangular Meshes. *ACM Transactions on Graphics*, 24(3):561–566, 2005. (Cited on pages 70 and 72.)

D. Kainmueller, T. Lange, and H. Lamecker. Shape Constrained Automatic Segmentation of the Liver based on a Heuristic Intensity Model. In B. van Ginneken,

T. Heimann, and M. Styner, editors, *3D Segmentation in the Clinic: A Grand Challenge*, pages 109–116, 2007. (Cited on pages 38, 39, 90 and 144.)

D. Kainmueller, H. Lamecker, S. Zachow, and H.-C. Hege. An Articulated Statistical Shape Model for Accurate Hip Joint Segmentation. In *International Conference of the IEEE Engineering in Medicine and Biology Society*, pages 6345–6351, 2009a. (Cited on page 137.)

D. Kainmueller, H. Lamecker, H. Seim, M. Zinser, and S. Zachow. Automatic Extraction of Mandibular Nerve and Bone from Cone-Beam CT Data. In G.-Z. Yang, D. Hawkes, D. Rueckert, A. Noble, and C. Taylor, editors, *MICCAI*, volume 5762 of *LNCS*, pages 76–83. Springer, 2009b. (Cited on pages 67, 68, 90, 117 and 119.)

D. Kainmueller, H. Lamecker, H. Seim, and S. Zachow. Multi-object Segmentation of Head Bones. *MIDAS Journal*, 2009c. (Cited on page 121.)

D. Kainmueller, H. Lamecker, and S. Zachow. Multi-object Segmentation with Coupled Deformable Models. *Annals of the British Machine Vision Association (BMVA)*, 5:1–10, 2009d. (Cited on pages 45 and 117.)

D. Kainmueller, H. Lamecker, H. Seim, S. Zachow, and H.-C. Hege. Improving Deformable Surface Meshes through Omni-directional Displacements and MRFs. In T. J. Navab, J. P. W. Pluim, and M. A. Viergever, editors, *MICCAI*, volume 6361 of *LNCS*, pages 227 – 234. Springer, 2010. (Cited on page 52.)

D. Kainmueller, H. Lamecker, B. Weber, M. Heller, C. Hege, and S. Zachow. Omnidirectional Displacements for Deformable Surfaces. *Medical Image Analysis*, 17 (4):429 – 441, 2013. (Cited on page 52.)

W. Kalender. *Computed Tomography*. Wiley, 2011. (Cited on pages 6 and 102.)

Y. Kang, K. Engelke, and W. Kalender. A New Accurate and Precise 3-D Segmentation Method for Skeletal Structures in Volumetric CT Data. *IEEE Transactions on Medical Imaging*, 22(5):586 –598, 2003. (Cited on page 119.)

A. Kaufman. An Algorithm for 3D Scan Conversion of Parametric Curves, Surfaces and Volumes. *Computer Graphics*, 21(4):171–179, 1987. (Cited on page 21.)

K. Khoshelham. Extending Generalized Hough Transform to Detect 3D Objects in Laser Range Data. In *ISPRS Workshop Laser Scanning*, page 206, 2007. (Cited on pages 23 and 32.)

C. Klapproth, A. Schiela, and P. Deuflhard. Fast Algorithms for the Simulation of Human Knee Joint Motion. In *Biomedical Engineering/765: Telehealth/766: Assistive Technologies*. ACTA Press, 2012. (Cited on page 70.)

J. Koenderink and A. van Doorn. Local Features of Smooth Shapes: Ridges and Courses. In *Proc. SPIE 2031, Geometric Methods in Computer Vision II*, pages 2–13, 1993. (Cited on page 64.)

D. Koller and N. Friedman. *Probabilistic Graphical Models: Principles and Techniques*. Adaptive Computation and Machine Learning Series. MIT Press, 2009. (Cited on page 54.)

N. Komodakis, G. Tziritas, and N. Paragios. Performance vs computational efficiency for optimizing single and dynamic MRFs: Setting the state of the art with primal-dual strategies. *Computer Vision and Image Understanding*, 112(1): 14–29, 2008. (Cited on pages 54, 56, 57, 63, 65 and 138.)

H. Lamecker, T. Lange, and M. Seebaß. A Statistical Shape Model for the Liver. In T. Dohi and R. Kikinis, editors, *MICCAI*, volume 2489 of *Lecture Notes in Computer Science*, pages 421–427. Springer, 2002. (Cited on pages 9, 30 and 91.)

H. Lamecker, M. Seebaß, H.-C. Hege, and P. Deuflhard. A 3D Statistical Shape Model of the Pelvic Bone for Segmentation. In J. Fitzpatrick and M. Sonka, editors, *Medical Imaging: Image Processing*, SPIE Conference Series, pages 1341–1351, 2004a. (Cited on pages 9, 30 and 100.)

H. Lamecker, T. Lange, and M. Seebaß. Segmentation of the Liver using a 3D Statistical Shape Model. Technical report, Zuse Institute Berlin, 2004b. (Cited on page 91.)

H. Lamecker, T. H. Wenckebach, H.-C. Hege, G. N. Duda, and M. O. Heller. Atlasbasierte 3D-Rekonstruktion des Beckens aus 2D-Projektionsbildern. In H. Handels, J. Ehrhardt, A. Horsch, H.-P. Meinzer, and T. Tolxdorff, editors, *Bildverarbeitung für die Medizin*, Informatik aktuell, pages 26–30. Springer-Verlag, 2006a. (Cited on page 149.)

H. Lamecker, S. Zachow, A. Wittmers, B. Weber, H.-C. Hege, B. Elsholtz, and M. Stiller. Automatic Segmentation of Mandibles in Low-dose CT-data. *International Journal of Computer Assisted Radiology and Surgery*, 1(1):393–395, 2006b. (Cited on pages 9 and 108.)

T. Langerak, U. Van der Heide, A. N. T. J. Kotte, M. Viergever, M. Van Vulpen, and J. P. W. Pluim. Label Fusion in Atlas-Based Segmentation Using a Selective and Iterative Method for Performance Level Estimation (SIMPLE). *IEEE Transactions on Medical Imaging*, 29(12):2000–2008, 2010. (Cited on page 73.)

K. Lee, R. K. Johnson, Y. Yin, A. Wahle, M. E. Olszewski, T. D. Scholz, and M. Sonka. Three-dimensional Thrombus Segmentation in Abdominal Aortic Aneurysms using Graph Search based on a Triangular Mesh. *Computers in Biology and Medicine*, 40(3):271 – 278, 2010. (Cited on page 42.)

G. E. Lewinnek, J. L. Lewis, R. Tarr, C. L. Compere, and J. R. Zimmerman. Dislocations after Total Hip Replacement. *Journal of Bone and Joint Surgery, American Volume*, 60:217–220, 1978. (Cited on pages 148 and 149.)

K. Li, S. Millington, X. Wu, D. Z. Chen, and M. Sonka. Simultaneous Segmentation of Multiple Closed Surfaces Using Optimal Graph Searching. In G. E. Christensen and M. Sonka, editors, *Information Processing in Medical Imaging*, volume 3565 of *LNCS*, pages 406–417. Springer, 2005. (Cited on page 44.)

K. Li, X. Wu, D. Z. Chen, and M. Sonka. Optimal Surface Segmentation in Volumetric Images-A Graph-Theoretic Approach. *IEEE Transactions on Pattern Analysis and Machine Intelligence*, 28(1):119–134, 2006. (Cited on pages 41, 42, 44, 45 and 60.)

S. Li. *Markov Random Field Modeling in Image Analysis*. Advances in Pattern Recognition. Springer-Verlag London, 2009. (Cited on pages 54 and 56.)

L. Liu, D. Raber, D. Nopachai, P. Commean, D. Sinacore, F. Prior, R. Pless, and T. Ju. Interactive Separation of Segmented Bones in CT Volumes Using Graph Cut. In D. N. Metaxas, L. Axel, G. Fichtinger, and G. Székely, editors, *MICCAI*, volume 5241 of *LNCS*, pages 296–304. Springer, 2008. (Cited on pages 11 and 119.)

X. Liu, D. Z. Chen, X. Wu, and M. Sonka. Optimal Graph Search based Image Segmentation for Objects with Complex Topologies. In J. P. W. Pluim and B. M. Dawant, editors, *Medical Imaging: Image Processing*, volume 7259, page 725915. SPIE, 2009. (Cited on page 44.)

W. Lorensen and H. Cline. Marching cubes: A high resolution 3d surface construction algorithm. *Computer Graphics*, 21(4):163–169, 1987. (Cited on page 21.)

J. B. A. Maintz and M. A. Viergever. A Survey of Medical Image Registration. *Medical Image Analysis*, 2(1):1–36, 1998. (Cited on page 74.)

S. Majumdar. *Advances in MRI of the Knee for Osteoarthritis*. World Scientific Publishing Company, Incorporated, 2010. (Cited on page 121.)

R. Malladi, J. Sethian, and B. Vemuri. Shape Modeling with Front Propagation: A Level Set Approach. *IEEE Transactions on Pattern Analysis and Machine Intelligence*, 17(2):158 –175, 1995. (Cited on page 11.)

T. McInerney and D. Terzopoulos. Deformable Models in Medical Image Analysis: A Survey. *Medical Image Analysis*, 1:91–108, 1996. (Cited on page 7.)

G. J. McLachlan and T. Krishnan. *The EM Algorithm and Extensions*. John Wiley & Sons, Inc., 2007. (Cited on page 35.)

A. Mharib, A. Ramli, S. Mashohor, and R. Mahmood. Survey on Liver CT Image Segmentation Methods. *Artificial Intelligence Review*, 37:83–95, 2012. (Cited on page 91.)

J. V. Miller, D. E. Breen, W. E. Lorensen, R. M. O'Bara, and M. J. Wozny. Geometrically Deformed Models: A Method for Extracting Closed Geometric Models form Volume Data. In *Proc. SIGGRAPH Conference on Computer Graphics and Interactive Techniques*, pages 217–226, New York, NY, USA, 1991. ACM. (Cited on page 39.)

J. Modersitzki. *Numerical Methods for Image Registration*. Oxford University Press, New York, 2004. (Cited on pages 18, 23, 74 and 76.)

J. Modersitzki. *FAIR: Flexible Algorithms for Image Registration*. SIAM, Philadelphia, 2009. (Cited on page 74.)

J. Montagnat and H. Delingette. Volumetric Medical Images Segmentation using Shape Constrained Deformable Models. In J. Troccaz, E. Grimson, and R. Mösges, editors, *CVRMed-MRCAS'97*, volume 1205 of *LNCS*, pages 13–22. Springer, 1997. (Cited on page 91.)

J. Montagnat, H. Delingette, and N. Ayache. A Review of Deformable Surfaces: Topology, Geometry and Deformation. *Image and Vision Computing*, 19:1023–1040, 2001. (Cited on pages 7 and 26.)

E. N. Mortensen and W. A. Barrett. Interactive Segmentation with Intelligent Scissors. *Graphical Models and Image Processing*, 60(5):349 – 384, 1998. (Cited on page 42.)

T. D. Nguyen, H. Lamecker, D. Kainmueller, and S. Zachow. Automatic Detection and Classification of Teeth in CT Data. In N. Ayache, H. Delingette, P. Golland, and K. Mori, editors, *MICCAI*, volume 7510 of *Lecture Notes in Computer Science*, pages 609–616. Springer Berlin Heidelberg, 2012. (Cited on page 114.)

W. Niessen, C. Bouma, K. Vincken, and M. Viergever. *Error Metrics for Quantitative Evalution of Medical Image Segmentation*, pages 275–284. Computational Imaging and Vision. Kluwer Academic Publishers, 2000. (Cited on page 83.)

X. Papademetris, D. Dione, L. Dobrucki, L. Staib, and A. Sinusas. Articulated Rigid Registration for Serial Lower-Limb Mouse Imaging. In *MICCAI*, volume 3750 of *LNCS*, pages 919–926. Springer, 2005. (Cited on page 154.)

J.-Y. Park, T. McInerney, D. Terzopoulos, and M.-H. Kim. A Non-self-intersecting Adaptive Deformable Surface for Complex Boundary Extraction from Volumetric Images. *Computers & Graphics*, 25(3):421–440, 2001. (Cited on page 53.)

R. R. Paulsen, J. A. Baerentzen, and R. Larsen. Markov Random Field Surface Reconstruction. *IEEE Transactions on Visualization and Computer Graphics*, 16:636–646, 2010. (Cited on page 54.)

V. Pekar, M. R. Kaus, C. Lorenz, S. Lobregt, R. Truyen, and J. Weese. Shape-model-based Adaptation of 3D Deformable Meshes for Segmentation of Medical Images. In M. Sonka and K. M. Hanson, editors, *Medical Imaging: Image Processing*, volume 4322 of *SPIE Conference Series*, pages 281–289, 2001. (Cited on page 38.)

V. Pekar, T. R. McNutt, and M. R. Kaus. Automated Model-based Organ Delineation for Radiotherapy Planning in Prostatic Region. *International Journal of Radiation Oncology Biology Physics*, 60(3):973–980, 2004. (Cited on page 4.)

V. Pekar, S. Allaire, J. Kim, and D. Jaffray. Head and Neck Auto-segmentation Challenge. *MIDAS Journal*, 2009. (Cited on page 121.)

J. Petersen, M. Nielsen, P. Lo, Z. Saghir, A. Dirksen, and M. de Bruijne. Optimal Graph Based Segmentation Using Flow Lines with Application to Airway Wall Segmentation. In G. Szkely and H. Hahn, editors, *Information Processing in Medical Imaging*, volume 6801 of *LNCS*, pages 49–60. Springer, 2011. (Cited on page 42.)

J. Pettersson, H. Knutsson, and M. Borga. Automatic Hip Bone Segmentation Using Non-Rigid Registration. In Y. Tang, S. Wang, G. Lorette, D. Yeung, and H. Yan, editors, *Proc. IEEE International Conference on Pattern Recognition (3)*, volume 3, pages 946–949. IEEE Computer Society, 2006. (Cited on page 101.)

D. Pham, C. Xu, J. Prince, et al. Current Methods in Medical Image Segmentation. *Annual review of biomedical engineering*, 2:315, 2000. (Cited on pages 2, 3 and 19.)

M. Poon, G. Hamarneh, and R. Abugharbieh. Live-Vessel: Extending Livewire for Simultaneous Extraction of Optimal Medial and Boundary Paths in Vascular Images. In N. Ayache, S. Ourselin, and A. Maeder, editors, *MICCAI*, volume 4792 of *LNCS*, pages 444–451. Springer, 2007. (Cited on page 114.)

S. Prohaska. *Skeleton-Based Visualization of Massive Voxel Objects with Network-Like Architecture*. Ph.d. thesis, Universität Potsdam, Mathematisch-Naturwissenschaftliche Fakultät (Jürgen Döllner), 2007. (Cited on page 47.)

R. Rothe. Zum Problem des Talwegs. *Sitzungsberichte der Berliner Mathematischen Gesellschaft*, 14:51–69, 1915. (Cited on page 64.)

S. Rueda, J. A. Gil, R. Pichery, and M. A. Raya. Automatic Segmentation of Jaw Tissues in CT Using Active Appearance Models and Semi-automatic Landmarking. In *MICCAI*, pages 167–174, 2006. (Cited on page 108.)

H. Ruppertshofen, C. Lorenz, S. Schmidt, P. Beyerlein, Z. Salah, G. Rose, and H. Schramm. Discriminative Generalized Hough Transform for Localization of Joints in the Lower Extremities. *Computer Science - Research and Development*, 26:97–105, 2011. (Cited on pages 33 and 92.)

C. Sammut and G. Webb. *Encyclopedia of Machine Learning*. Springer reference. Springer, 2011. (Cited on page 86.)

M. Sato, I. Bitter, M. Bende, A. Kaufman, and M. Nakajima. TEASAR: Tree-structure Extraction Algorithm for Accurate and Robust Skeletons. In B. A. Barsky, Y. Shinagawa, and W. Wang, editors, *Proc. Pacific Conference on Computer Graphics and Application*, pages 281–289, Los Alamitos, CA, 2000. IEEE. (Cited on page 109.)

A. Schenk, G. Prause, and H.-O. Peitgen. Efficient Semiautomatic Segmentation of 3D Objects in Medical Images. In S. Delp, A. DiGoia, and B. Jaramaz, editors, *MICCAI*, volume 1935 of *Lecture Notes in Computer Science*, pages 71–131. Springer Berlin / Heidelberg, 2000. (Cited on pages 11 and 42.)

P. J. Schneider and D. H. Eberly. *Geometric Tools for Computer Graphics*. Morgan Kaufmann Pub, 2003. (Cited on pages 19 and 20.)

A. Schramm, M. Rücker, N. Sakka, Schön, J. Düker, and N.-C. Gellrich. The Use of Cone Beam CT in Cranio-maxillofacial Surgery. In H. Lemke, K. Inamura, K. Doi, M. Vannier, and A. Farman, editors, *Computer Assisted Radiology and Surgery*, volume 1281 of *Int. Congress Series*, pages 1200–1204. Elsevier, 2005. (Cited on pages 90 and 108.)

H. Seim, D. Kainmueller, M. Heller, H. Lamecker, S. Zachow, and H.-C. Hege. Automatic Segmentation of the Pelvic Bones from CT Data Based on a Statistical Shape Model. In *Eurographics Workshop on Visual Computing for Biomedicine*, pages 93–100, 2008. (Cited on page 90.)

H. Seim, D. Kainmueller, M. Heller, S. Zachow, and H.-C. Hege. Automatic Extraction of Anatomical Landmarks from Medical Image Data: An Evaluation of Different Methods. In *International Symposium on Biomedical Imaging*, pages 538–541, 2009. (Cited on page 147.)

H. Seim, D. Kainmueller, H. Lamecker, M. Bindernagel, J. Malinowski, and S. Zachow. Model-based Auto-Segmentation of Knee Bones and Cartilage in MRI

Data. In B. van Ginneken, K. Murphy, T. Heimann, V. Pekar, and X. Deng, editors, *Proc. MICCAI Workshop Medical Image Analysis for the Clinic: A Grand Challenge*, pages 215 – 223, 2010. (Cited on page 117.)

A. Sen and M. Srivastava. *Regression Analysis: Theory, Methods, and Applications*. Springer Texts in Statistics. Springer, 1990. (Cited on page 85.)

H. Shim, S. Chang, C. Tao, J.-H. Wang, C. K. Kwoh, and K. T. Bae. Knee Cartilage: Efficient and Reproducible Segmentation on High-spatial-resolution MR Images with the Semiautomated Graph-cut Algorithm Method. *Radiology*, 251(2):548–56, 2009. (Cited on page 4.)

A. R. Smith. A Pixel Is Not A Little Square, A Pixel Is Not A Little Square, A Pixel Is Not A Little Square! (And a Voxel is Not a Little Cube). Technical report, Technical Memo 6, Microsoft Research, 1995. (Cited on page 18.)

M. Sonka and J. M. Fitzpatrick, editors. *Handbook of Medical Imaging: Medical Image Processing and Analysis*. Press Monographs. SPIE Press, 2000. (Cited on pages 2 and 5.)

D. Stalling and H.-C. Hege. Intelligent Scissors for Medical Image Segmentation. In B. Arnolds, H. Müller, D. Saupe, and T. Tolxdorff, editors, *Bildverarbeitung für die Medizin*, pages 32 – 36, 1996. (Cited on page 42.)

D. Stalling, M. Westerhoff, and H.-C. Hege. Amira: A Highly Interactive System for Visual Data Analysis. In C. D. Hansen and C. R. Johnson, editors, *The Visualization Handbook*, chapter 38, pages 749–767. Elsevier, 2005. (Cited on page 2.)

W. Stein, S. Hassfeld, and J. Muhling. Tracing of Thin Tubular Structures in Computer Tomographic Data. *Computer Aided Surgery*, 3:83–88, 1998. (Cited on page 108.)

V. Surazhsky and C. Gotsman. Explicit Surface Remeshing. In *Proc. Eurographics Symposium on Geometry Processing*, pages 20–30. Eurographics Association, 2003. (Cited on page 142.)

K. Tönnies. *Guide to Medical Image Analysis*. Advances in Computer Vision and Pattern Recognition. Springer London, 2012. (Cited on page 5.)

T. Torsney-Weir, A. Saad, T. Moller, H. Hege, B. Weber, J. Verbavatz, and S. Bergner. Tuner: Principled Parameter Finding for Image Segmentation Algorithms Using Visual Response Surface Exploration. *IEEE Transactions on Visualization and Computer Graphics*, 17(12):1892 –1901, 2011. (Cited on page 87.)

J. Udupa and G. Herman, editors. *3D Imaging in Medicine*. CRC Press INC., 2000. (Cited on pages 2 and 3.)

B. van Ginneken, T. Heimann, and M. Styner. 3D Segmentation in the Clinic: A Grand Challenge. In B. van Ginneken, T. Heimann, and M. Styner, editors, *Proc. MICCAI Workshop 3D Segmentation in the Clinic: A Grand Challenge*, pages 7–15, 2007. (Cited on pages 4, 82, 83, 85, 92 and 96.)

S. Vasilache and K. Najarian. Automated Bone Segmentation from Pelvic CT Images. In *IEEE International Conference on Bioinformatics and Biomedicine Workshops*, pages 41 –47, 2008. (Cited on page 101.)

A. Viterbi. Error Bounds for Convolutional Codes and an Asymptotically Optimum Decoding Algorithm. *IEEE Transactions on Information Theory*, 13(2):260 –269, 1967. (Cited on page 42.)

B. Weber. Merkmalskurven auf triangulierten Oberflächen. Master's thesis, Department of Mathematics and Computer Science, FU-Berlin, 2008. Advised by Konrad Polthier and Steffen Prohaska. (Cited on pages 64 and 65.)

J. Weese, M. Kaus, C. Lorenz, S. Lobregt, R. Truyen, and V. Pekar. Shape Constrained Deformable Models for 3D Medical Image Segmentation. In M. Insana and R. Leahy, editors, *Information Processing in Medical Imaging*, volume 2082 of *LNCS*, pages 380–387. Springer, 2001. (Cited on page 91.)

J. Weickert, B. M. ter Haar Romeny, and M. A. Viergever. Efficient and Reliable Schemes for Nonlinear Diffusion Filtering. *IEEE Transactions on Image Processing*, 7(3):398–410, 1998. (Cited on page 92.)

W. C. Wong and A. C. Chung. Probabilistic Vessel Axis Tracing and its Application to Vessel Segmentation with Stream Surfaces and Minimum Cost Paths. *Medical Image Analysis*, 11(6):567–587, 2007. (Cited on page 114.)

S. Wörz and K. Rohr. Localization of Anatomical Point Landmarks in 3D Medical Images by Fitting 3D Parametric Intensity Models. *Medical Image Analysis*, 10 (1):41–58, 2006. (Cited on page 149.)

J. Wu, P. Davuluri, K. Ward, C. Cockrell, R. Hobson, and K. Najarian. A New Hierarchical Method for Multi-level Segmentation of Bone in Pelvic CT Scans. In *International Conference of the IEEE Engineering in Medicine and Biology Society*, pages 3399 –3402, 2011. (Cited on page 101.)

C. Xu, D. L. Pham, and J. L. Prince. *Handbook of Medical Imaging. Vol.2 Medical Image Processing and Analysis*, volume 2, chapter Image Segmentation using Deformable Models, pages 129–174. SPIE, 2000. (Cited on page 7.)

H. T. Yau, Y. K. Lin, L. S. Tsou, and C. Y. Lee. An Adaptive Region Growing Method to Segment Inferior Alveolar Nerve Canal from 3D Medical Images for Dental Implant Surgery. *Computer-Aided Design and Applications*, 5(5):743–752, 2008. (Cited on page 108.)

Y. Yin, X. Zhang, and M. Sonka. Optimal Multi-Object Multi-Surface Graph Search Segmentation: Full-Joint Cartilage Delineation in 3D. In S. McKenna and J. Hoey, editors, *Medical Image Understanding and Analysis*, pages 104–108, 2008. (Cited on page 44.)

Y. Yin, X. Zhang, D. D. Anderson, T. D. Brown, C. V. Hofwegen, and M. Sonka. Simultaneous Segmentation of the Bone and Bartilage Surfaces of a Knee Joint in 3D. In *Medical Imaging: Image Processing*, SPIE Conference Series, pages 72591O–72591O–9, 2009. (Cited on page 122.)

Y. Yin, X. Zhang, R. Williams, X. Wu, D. D. Anderson, and M. Sonka. LOGISMOS - Layered Optimal Graph Image Segmentation of Multiple Objects and Surfaces: Cartilage Segmentation in the Knee Joint. *IEEE Transactions on Medical Imaging*, 29(12):2023–2037, 2010. (Cited on pages 23 and 42.)

F. Yokota, T. Okada, M. Takao, N. Sugano, Y. Tada, and Y. Sato. Automated Segmentation of the Femur and Pelvis from 3D CT Data of Diseased Hip Using Hierarchical Statistical Shape Model of Joint Structure. In G.-Z. Yang, D. Hawkes, D. Rueckert, A. Noble, and C. Taylor, editors, *MICCAI*, volume 5762 of *Lecture Notes in Computer Science*, pages 811–818. Springer Berlin / Heidelberg, 2009. (Cited on pages 100 and 119.)

S. Zachow, H. Lamecker, B. Elsholtz, and M. Stiller. Is the Course of the Mandibular Nerve Deducible from the Shape of the Mandible? *International Journal of Computer Assisted Radiology and Surgery*, 1(1):415–417, 2006. (Cited on page 108.)

X. Zhang, J. Tian, K. Deng, Y. Wu, and X. Li. Automatic Liver Segmentation Using a Statistical Shape Model With Optimal Surface Detection. *IEEE Transactions on Biomedical Engineering*, 57(10):2622–2626, 2010. (Cited on page 42.)

R. A. Zoroofi, Y. Sato, T. Sasama, T. Nishii, N. Sugano, K. Yonenobu, H. Yoshikawa, T. Ochi, and S. Tamura. Automated Segmentation of Acetabulum and Femoral Head from 3-d CT Images. *IEEE Transactions on Information Technology in Biomedicine*, 7(4):329–343, 2003. (Cited on page 119.)

Aktuelle Forschung Medizintechnik

Herausgeber:

Prof. Dr. Thorsten M. Buzug

Institut für Medizintechnik, Universität zu Lübeck

Themen
Werke aus folgenden Themengebieten werden gerne in die Reihe aufgenommen: Biomedizinische Mikro- und Nanosysteme, Elektromedizin, biomedizinische Mess- und Sensortechnik, Monitoring, Lasertechnik, Robotik, minimalinvasive Chirurgie, integrierte OP-Systeme, bildgebende Verfahren, digitale Bildverarbeitung und Visualisierung, Kommunikations- und Informationssysteme, Telemedizin, eHealth und wissensbasierte Systeme, Biosignalverarbeitung, Modellierung und Simulation, Biomechanik, aktive und passive Implantate, Tissue Engineering, Neuroprothetik, Dosimetrie, Strahlenschutz, Strahlentherapie.

Autorinnen und Autoren
Autoren der Reihe sind in der Regel junge Promovierte und Habilitierte, die exzellente Abschlussarbeiten verfasst haben.

Leserschaft
Die Reihe wendet sich einerseits an Studierende, Promovenden und Habilitanden aus den Bereichen Medizintechnik, Medizinische Ingenieurwissenschaft, Medizinische Physik, Medizinische Informatik oder ähnlicher Richtungen. Andererseits stellt die Reihe aktuelle Arbeiten aus einem sich schnell entwickelnden Feld dar, so dass auch Wissenschaftlerinnen und Wissenschaftler sowie Entwicklerinnen und Entwickler an Universitäten, in außeruniversitären Forschungseinrichtungen und der Industrie von den ausgewählten Arbeiten in innovativen Gebieten der Medizintechnik profitieren werden.

Begutachtungsprozess
Die Qualitätssicherung erfolgt in drei Schritten. Zunächst werden nur Arbeiten angenommen die mindestens magna cum laude bewertet sind. Im zweiten Schritt wird ein Mitglied des Editorial Boards die Annahme oder Ablehnung des Werkes empfehlen. Im letzten Schritt wird der Reihenherausgeber über die Annahme oder Ablehnung entscheiden sowie Änderungen in der Druckfassung empfehlen. Die Koordination übernimmt der Reihenherausgeber.

Kontakt
Prof. Dr. Thorsten M. Buzug
Institut für Medizintechnik
Universität zu Lübeck
Ratzeburger Allee 160
23538 Lübeck, Germany

Tel.: +49 (0) 451 / 500-5400
Fax: +49 (0) 451 / 500-5403
E-Mail: buzug@imt.uni-luebeck.de
Web: http://www.imt.uni-luebeck.de

Stand: Januar 2014. Änderungen vorbehalten.
Erhältlich im Buchhandel oder beim Verlag.

Abraham-Lincoln-Straße 46
D-65189 Wiesbaden
Tel. +49 (0)6221. 345 - 4301
www.springer-vieweg.de

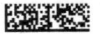